Leadership and Challenges in Medical Physics: A Strategic and Robust Approach

A EUTEMPE network book

IPEM–IOP Series in Physics and Engineering in Medicine and Biology

About the Series

Series in Physics and Engineering in Medicine and Biology will allow IPEM to enhance its mission to 'advance physics and engineering applied to medicine and biology for the public good.'

Focussing on key areas including, but not limited to:

- clinical engineering
- diagnostic radiology
- informatics and computing
- magnetic resonance imaging
- nuclear medicine
- physiological measurement
- radiation protection
- radiotherapy
- rehabilitation engineering
- ultrasound and non-ionising radiation.

A number of IPEM–IOP titles are published as part of the EUTEMPE Network Series for Medical Physics Experts.

Leadership and Challenges in Medical Physics: A Strategic and Robust Approach

A EUTEMPE network book

Carmel J Caruana
Medical Physics Department, Faculty of Health Sciences,
University of Malta, Msida, MSD 2080, Malta

IOP Publishing, Bristol, UK

ISBN 978-0-7503-1395-7 (ebook)
ISBN 978-0-7503-1396-4 (print)
ISBN 978-0-7503-1966-9 (myPrint)
ISBN 978-0-7503-1397-1 (mobi)

DOI 10.1088/978-0-7503-1395-7

Version: 20200301

IOP ebooks

British Library Cataloguing-in-Publication Data: A catalogue record for this book is available from the British Library.

Published by IOP Publishing, wholly owned by The Institute of Physics, London

IOP Publishing, Temple Circus, Temple Way, Bristol, BS1 6HG, UK

US Office: IOP Publishing, Inc., 190 North Independence Mall West, Suite 601, Philadelphia, PA 19106, USA

I would like to dedicate this work to all Medical Physicists worldwide who strive daily to improve healthcare even when not always shown the appreciation they so rightly deserve.

I dedicate it to the many past and present colleagues at EFOMP and IMPCB and thank them for the many stimulating discussions over meals and coffee breaks.

I also dedicate it to the past participants of the EFOMP-EUTEMPE module 'Leadership and challenges in Medical Physics' who have been a continuous source of inspiration with their enthusiasm, feedback and commitment to patients and profession.

Last but not least, I dedicate this work to my partner Michaela and thank her for her intelligent conversation which helped shape my thinking and to my daughters Antonella, Jane and Sara for their company and for their patience in those instances when their dad was lost in strategic thinking mode—they too have to accept strategic planning as part of modern professional life.

Professor Carmel J. Caruana, PhD FIPEM
Head, Medical Physics department, Faculty of Health Sciences,
University of Malta

Contents

Foreword

As a group, Medical Physicists are excellent scientists and healthcare professionals. Indeed, our exceptional education and training in physics, mathematics, medical devices, radiation and other physical agents and information technology has made us what we are today—a highly successful profession that has changed the face of healthcare. Every day we save lives by ensuring the clinically-effective, safe and efficient use of sophisticated diagnostic and therapeutic devices which are marvels of modern technology. As a profession we are mindful of the importance of what we do for the well-being of patients and others. We take our work with the required probity, commitment and devotion. Our education and training curricula are up there with the very best.

On the other hand, our education and training programmes provide little experience in how to deal with the real world issues facing us when we move from the relative security of our academic physics departments to the realities of modern, large, complex healthcare organizations. In a highly-competitive world of austerity economics, hospitals have in many instances been converted from what they should be, that is, humane, equitable, patient-oriented enterprises, into shopping malls for healthcare services. The healthcare industry, whilst extolling the virtues of inter-professional teamwork, has been reduced from a richly stimulating, intellectually satisfying multi-professional environment to a battleground where 'role-development' is not based on innovative thinking and a soul-fulfilling endeavour towards improved patient services, nor indeed mutual inter-professional respect, but in some cases nothing else but a work environment plagued with covert attempts at poaching the role of others by any means possible, however unethical. The challenges to the profession arising from these negative characteristics of present healthcare organizations need to be acknowledged and countered by the profession if we are to be significant players in the healthcare environment of the future. With this book, I hope to provoke much needed discussion and debate and to encourage more Medical Physicists to devote some of their quality time to these and other highly relevant issues. These challenges need to be addressed in the same way that we address all our Medical Physics problems: we need to analyse them, research them and solve them. To do this, we need strategic and robust leaders who are well-prepared to take on these tasks, who do not only aim to preserve the gains of the past but can push the profession to new heights. However, such leaders need to be educated and trained and the resources for this simply do not exist. I am hoping that this book will be a first solid attempt at addressing this lacuna in our education and training.

This book will help you become a strategic and robust leader. However, *please keep in mind that no book will turn you into a leader—ultimately YOU HAVE TO MAKE YOURSELF ONE.* We are a highly intelligent group, we have excelled as clinical scientists—I am very confident that we can excel in leadership too provided we give it the quality attention it deserves.

Professor Carmel J Caruana, PhD FIPEM
Head, Medical Physics department, Faculty of Health Sciences,
University of Malta

Acknowledgements

I would like to thank all National, European and International Medical Physics colleagues who with their insightful discussions helped form my thinking over the years. I would also like to thank past students and participants of my Medical Physics leadership courses for their challenging questions; each such question was an opportunity to further develop my own thinking and leadership skills. I consider this text not solely my own—it is also the cumulative total of the wisdom and experiences of many others.

Author biography

Carmel J Caruana

Professor Carmel J Caruana, PhD FIPEM has a BSc in Physics and Mathematics from the University of Malta, an MSc in Applied Radiation Physics from the University of Birmingham, UK and a PhD from Charles University, Prague. He is head of Medical Physics at the University of Malta and teaches diagnostic and interventional radiology, protection from radiation and other physical agents, and medical devices to medical physicists, radiologists, radiographers, nurses, physicians and applied biomedical scientists. In his home country, he drove the development of the profession from the time when it was totally unknown to the point when it was recognized legally as a healthcare profession. However, his greatest contributions to Medical Physics have been his education and training, role and professional issues initiatives at the European and International level. He is past chairperson of the education and training committee of the European Federation of Organizations for Medical Physics (EFOMP), lead author of the role and education and training chapters of the EU sponsored document 'European Guidelines on the Medical Physics Expert' and several EFOMP policy statements. He represented EFOMP on the MEDRAPET, ENETRAP and EUTEMPE projects. He set up the EFOMP School for Medical Physics Experts. He is the main author and organizer of the EFOMP-EUTEMPE module on leadership in Medical Physics. He was Associate Editor for educational and training and professional issues for *Physica Medica* (*European Journal of Medical Physics*) and is a member of the Editorial Advisory Board of the IOPP-IPEM Physics and Engineering in Medicine ebook series. He was one of the initial drivers of the International Medical Physics Certification Board.

List of acronyms

AAPM	American Association of Physicists in Medicine
CPD	Continuous professional development
D&IR	Diagnostic and interventional radiology
EFOMP	European Federation of Organizations for Medical Physics
EUTEMPE	EUropean Training and Education for Medical Physics Experts
IMPCB	International Medical Physics Certification Board
IOMP	International Organization for Medical Physics
KPI	Key performance indicators
MPE	Medical Physics expert
MPP	Medical Physics professionals
NM	Nuclear medicine
PESTLE	Political Environmental Socio-psychological Techno-scientific LEgal
QC	Quality control
RO	Radiation oncology
RPE	Radiation protection expert
SIG	Special interest group
SWOT	Strengths, weaknesses, opportunities, threats
TG	Task group

Chapter 1

What is strategic and robust leadership, and why is it critical for Medical Physics in the present environment?

Learning outcomes

By the end of the chapter the reader will be able to:

- distinguish between leadership and management;
- define strategic and robust leadership and explain its importance for Medical Physics today;
- distinguish between different types of Medical Physics groups and teams that one could lead;
- define the terms mission, vision, strategic plan;
- define the different types of intelligences and discuss their importance for leadership;
- understand how one can prepare oneself for leadership and that leadership is a personal journey;

1.1 Leadership versus management: there is a difference!

Management and leadership are not equivalent. *Leadership is the process of influencing and motivating others to agree on and work towards an exciting shared future vision; there is a focus on inspiring others and creating shared organizational culture and values.* Managers are employed to get things done by making sure administrative tasks such as planning, organizing, budgeting, quality controlling, staffing and problem-solving are carried out effectively and without unnecessary waste of resources—the role of a manager is closer to that of an executive officer. The two roles—leadership and management—are both essential but they are not the same. In practice, owing to insufficient human resources leaders often also need to

doi:10.1088/978-0-7503-1395-7ch1

take on managerial roles and *vice versa*. However, one cannot emphasize enough that management devoid of a future positive common vision to work towards and to dream of leads to a soulless, monotonous, demotivating and ultimately failing work environment. Very importantly, one needs to keep in mind that leadership is not administration and certainly leadership is not about being a boss.

1.2 Improving patient services by leading Medical Physics groups and teams

Leadership in Medical Physics is important because without good leadership clinical or research teams fail to deliver and the profession as a group would not develop locally, nationally, regionally or internationally. On account of the unique services provided by Medical Physicists such failures would ultimately result in a degradation of the effectiveness, safety and efficiency of patient services.

1.3 What type of Medical Physics group or team are you or would you be leading?

As a Medical Physics leader you would be leading a specific group or team of Medical Physicists (and sometimes multidisciplinary teams) and your first task as a leader would be to recognize the type and characteristics of the group or team you would be leading.

It is important to distinguish between the terms 'group' and 'team'. The term group refers to a set of individuals who have some common characteristic e.g. the group of Medical Physicists in a country. On the other hand 'team' refers to a small group with a high degree of interaction between the individual members and sharing a common purpose. A team is more than the sum of a set of individuals as the degree of synergy between the members makes possible the attainment of objectives which would not be possible by the individual members on their own. In the literature the two terms are often used interchangeably and keeping a clear distinction between the two is often difficult.

Such Medical Physics groups/teams could be:

- a new team of Medical Physics trainees;
- an ongoing special interest group (SIG) dedicated to the ongoing development of a specific area of professional practice;
- a time-limited team set up for a specific well-defined objective—often called a task group (TG);
- a team of Medical Physicists within an independent Medical Physics department (clinical/academic/combined);
- a small team of Medical Physicists within a diagnostic and interventional radiology/radiation oncology/nuclear medicine department managed by physicians or radiographers;
- the members of a Medical Physics professional organization;
- a multidisciplinary team set up for a specific task;
- increasingly, virtual teams with members geographically distributed in different countries and time zones.

It is very important to reflect on the group/team that you would be leading as different types of groups/teams have different objectives (e.g. research, clinical, professional, educational, policy-making, political) and characteristics (e.g. size, level of homogeneity or otherwise, local/national/regional/international, level of motivation, degree of collaboration required to achieve purpose, role assignment, degree of member specialization). Different types of groups/teams require different styles of leadership and motivational strategies.

1.3.1 Reflection/discussion point

Consider the above types of groups/teams. Different groups/teams have different purpose: research, clinical, professional, educational, policy-making, political. All types are important for the profession to develop and flourish. Which types of groups/teams are you leading or do you aspire to lead in the future? Which type fits with your own views on leadership and personal characteristics? Consider which types are very much necessary at this point in time in the development of the profession in your own country—these are the ones to aim for.

1.4 Leaders lead people

It is very important to always keep in mind that even when one is elected as a leader of a Medical Physics department, professional organization (local, national, regional, international) one is ultimately there to lead *people*, i.e. the group of professionals, members of a particular department or organization. It is important to constantly remind oneself of this important *raison d'être* of leadership as sometimes the organizational challenges facing the department or organization *per se* (e.g. issues of finance, recognition, competition by a parallel organization, intra/inter-departmental and intra-organizational politics) tend to take over, and the interests of the members end up on the back-burner.

1.5 Leadership does not happen in a vacuum but in specific environments

It is important to characterise also the type of physical and socio-political organizational environment in which the group or team resides. For example, what is the level of the healthcare organization[1]? What physical size? Is it a modern building or

[1] The four levels of healthcare organizations are: (a) PRIMARY care is the level with which most patients are familiar, it is the first stop when patients experience new symptoms or experience minor trauma. Primary care should also function as a coordinating centre for care. (b) SECONDARY care includes specialists. Primary care providers may refer patients to specialists found in secondary care centres. Specialists are very knowledgeable on specific body systems, body region, disease or condition. (c) TERTIARY care centres are hospitals which provide highly specialized equipment and expertise such as haemodialysis, coronary artery bypass surgery, neurosurgery, plastic surgery or severe burn treatment. (d) QUATERNARY care is an extension of tertiary care which includes experimental medicine and highly exceptional surgical procedures. Few hospitals offer quaternary care. Most Medical Physicists are to be found in tertiary and quaternary healthcare organizations whilst primary and secondary facilities are serviced by external Medical Physics consultants.

is it in decay? In the case of professional organizations: is it local, national, regional, international; based on individual membership or is it a federation of associations? Is the socio-political environment in which the healthcare organization resides and indeed its own internal organizational environment democratic or autocratic? Is the organizational structure flat or pyramidal?

1.6 Collective leadership of Medical Physics associations by boards

Leadership of members of a Medical Physics association is the responsibility of the board of that association. The board (consisting of a chairperson, officers and members) provides *collective leadership* to the members of the association/ organization. In the rest of the book the terms leader and leadership will apply to both the leadership provided by an individual leader and the collective responsibility provided by a board.

1.7 Some basic definitions: mission, vision, strategic plan

We all have a generic idea of what leadership is and different definitions abound in the literature (just Google it!). We need to establish a definition which would be suitable for our profession. However, before we do that we need some terminology. Here goes:

- the *MISSION of a group/team* is basically a statement which answers the question: what specific *unique benefits does the group/team provide and to whom*? (who are the direct and indirect clients of the group: patients, other healthcare professionals, hospital management?);
- the *VISION of the group/team* is a statement answering the question: *keeping in mind the mission* what *state* does the *group/team* aspire to by a specific point in time in the *future*?
- a *STRATEGIC PLAN* is a document detailing a process for *bridging the gap between the present state of the group/team and the desired future vision.*

These concepts will be considered in more detail in chapter 2.

1.8 Definition of strategic leadership

We define STRATEGIC LEADERSHIP as the ability to guide a specific group/team in developing an appropriate vision, to persuade the members of the importance of that vision, to facilitate the development of a strategic plan to achieve that vision and to be able to motivate them to work collaboratively, willingly and effectively towards ensuring that the vision is achieved.

A strategic leader is a developer, promoter and facilitator of vision. A strategic leader is a person who inspires their colleagues by their ability to create an exciting yet practical realistic vision, their passion for that vision, the personal energy they input into the achievement of that vision and their ability to motivate others to share and work together towards the achievement of that vision.

1.9 Definition of robust leadership

We characterise leaders as being ROBUST when they are capable of maintaining focus on vision and *adapting their mental perspectives and psychological behaviours such that the vision is achieved irrespective of environmental change, conditions and stressors (e.g. political, economic, socio-psychological, techno-scientific, legal)*. It is easy to do well when environments are favourable and supportive but difficult times test a leader's robustness. *Difficult times are challenging but they are also an opportunity for a leader to develop their robustness and their stature as a leader among their peers.* Maintaining focus is often not easy, particularly in the hospital environment, as there are often too many demands competing for one's attention. Maintaining focus requires self-discipline and an ability to apportion your time, energy and resources wisely to what is really important and to do so in an ongoing consistent manner over the years. Robust leaders need to be psychologically tough, resilient, bold, and respectful of others without being timid—robust leaders may bend but not break!

1.9.1 Reflection/discussion point

Try to think of instances in your life when you were part of a team with an assigned leader preferably a Medical Physics one (however, it could be any type really, from a bunch of friends organizing a party to a football team):

- did the team have a clear mission?
- did the leaders develop a clear vision and was that vision communicated to the individual members well?
- which teams achieved their vision, which teams did not?
- analyze the character and behaviours of the leader in each case. What were the differences between the leaders of successful and failing teams?
- what could the failed leaders have done differently to make them more successful as leaders?
- when leadership failed was it a question of lack of vision, strategic thinking, insufficient robustness on the part of the leaders?

1.10 Strategic and robust leadership has become critical for Medical Physics today

In today's world, being a good scientist is simply not enough to survive and thrive. We live in a rapidly changing world dominated by austerity-based economics, reduced budgets and sometimes uninspiring political leaders and role models. In such a socio-political-economic environment actual healthcare service quality often leaves much to be desired. Who cares as long as the patients and newspapers don't detect it and it doesn't get splashed over the front pages and we lose votes in the process? In such environments our intrinsic, precious professional values of objectivity, quality, accuracy, precision and safety are often paid mere lip service. In such a relatively hostile environment strategic and robust leaders are important for all professions. This is true both for relatively large politically powerful professions like nursing or small yet still politically powerful professions like radiology. However, they become crucial in the case of small politically not-so-strong professions like Medical Physics. Strategic leadership is one where leadership decisions are taken not on the basis of intuition or personality or on the basis of spur of the moment decisions or a crisis management approach but on carefully well-thought-out research-based strategic plans based on clear and relevant mission and vision. Strategic leadership ensures the mission is put in practice and the vision achieved. It has been argued that the fostering of future strategic leaders has become so important that training should start at the level of the masters in Medical Physics [1]. Let us consider two main reasons why strategic and robust leaders are required in Medical Physics today: conflicting world views and unrestrained austerity economics and immoderate commoditization.

1.11 Galilean versus Darwinian world views

As Medical Physicists we are steeped in the values of Galileo. We believe that society moves forward if its members are taught to revere truth and objectivity and to negotiate in a principled manner (i.e. based on objective principles as opposed to subjective self-interest). As scientists we are rewarded for our objectivity and disparaged if we exhibit subjectivity. As good scientists we are taught (and rightly so!) that to produce good science we need to *doubt ourselves in a systematic manner*:

- am I taking the right measurements?
- am I using the proper instrument?
- are my measurements accurate?
- is the precision level sufficient?
- is the equation correct?
- does the equation apply in these particular circumstances?
- am I permitting my own wishes to cloud my judgement?

Yet we live in a post-truth world where 'alternative facts' based on subjective emotions and personal interests and forthright bluff have become common and in some environments the order of the day. Even the healthcare environment has not been spared. If you doubt yourself, you are considered weak and you are expected to appear confident even when you are actually not. You are expected to make statements which sound authoritative even when you do not have the knowledge or evidence to support them. If you can 'sell it' you are considered successful even if it means being insincere. You are considered successful if you can peddle yourself even if it means faking knowledge and skills. Honesty of process and ethical principles are sometimes sacrificed in the belief that outcomes reign supreme as long as there is a thin veneer of authenticity which hides the less desirable truth from the world outside. Hospitals have been turned into giant competition based shopping malls for healthcare services. The inter-professional healthcare milieu has been turned into a battleground where role-development is not based on creative thinking, improved patient services and mutual respect between the professions but an environment dominated by the poaching of the role of others by any means possible, however unethical. Even research targeted towards the development of the professions has become political and has to be critically evaluated as results may be based on false premises and dodgy methodologies. In such Darwinian survival-of-the-fittest scenarios, battles are not won by the ablest and most intelligent but by those who are readiest to ride the tide of the current local political (with a big and small 'p') environment even though such a Darwinian adaptation leads to loss of personal integrity and a degradation of patient service.

1.12 Austerity economics and immoderate commoditization

AUSTERITY ECONOMICS aims to reduce government budget deficits through spending cuts, tax increases, or a combination of both. The first targets of spending cuts are often public education, social services and of course public healthcare. COMMODITIZATION refers to the process by which parts of services offered by high level professionals such as Medical Physicists and radiologists become cheaper to provide by establishing uniformly rigid written protocols which permit such services to be carried out by lower level (and hence cheaper) professionals (e.g. basic chest reporting done by specially trained radiographers instead of fully-trained radiologists; daily and weekly quality control (QC) carried out by Medical Physics assistants [2] or radiographers). In modern societies, the ability to commoditize anything is seen as a benefit to all, and opens up higher level human resources that can be put to better use for service development and innovation e.g. Medical Physicists having more time to focus on higher order tasks as opposed to spending time on routine simple QC tasks. However, if not managed well, commoditization would ultimately pose a threat to the higher level professions [2] and ultimately quality of patient service e.g. radiology department managers under pressure to reduce costs opting to save money by employing a lower paid worker initially employed to do routine QC who is then also tasked to do advanced QC even though he/she is not capable—with the result that patient safety degrades; or diagnostic radiographers being employed to read chest radiographs without taking the necessary precautions to ensure that challenging cases are automatically transferred to chest radiologists, leading to increases in the number of misdiagnoses. Such commoditization has other effects: often problems of equity[2] arise as rich patients would be able to afford fully-qualified radiologists reducing the risk of misdiagnosis whilst other patients would not. Questions arise: in the case of public institutions, who would decide whether a patient is directed towards a fully-qualified radiologist as opposed to a radiographer and on what criteria? There are clearly issues of social justice involved. Again, patients attending large radiation oncology centres with well-developed and fully staffed Medical Physics departments get accurate and precise treatment; patients where no such departments exist have a higher rate of mortality [3–5]. This is particularly common in those departments when

[2] One should distinguish between *equity and equality*. Equity refers to differences in health of different patients and populations arising from differences in socioeconomic status, such differences are remediable. Equality on the other hand is not possible as there are factors which would impede its attainment which are totally beyond our control e.g. genetic factors.

departmental managers are not sufficiently informed about the full range of services offered by radiologists and Medical Physicists.

1.12.1 Reflection/discussion point

Consider references [3–5]. In article [3] we find the statement: 'The accident resulted from a convergence of different human errors. There were three main sources of error, discussed below. These were related to inaccurate "in-house" calculation of the MU, improper use of portal imaging and inadequate use of dynamic wedges for IMRT'. The abstract of reference [4] includes the following: 'The treatments were interrupted in this hospital for several months, to organize a new management of the department, based on quality and safety. The preventive actions to avoid such an accident necessitate to evaluate standard and innovative treatments, to develop an internal and external quality control program.' Reference [5] about the same incident also mentions calibration errors. Would such mistakes have occurred if there had been an operational Medical Physics department?

1.12.2 Reflection/discussion point

'Overdosing of patients in diagnostic and interventional radiology is often a daily affair where there are no Medical Physicists within diagnostic and interventional radiology departments, yet it rarely makes the headlines because the *individual* dose is seldom so high so as to produce immediate visible acute effects. However, the *population* doses are enormous owing to the millions of imaging studies carried out daily. This population overdosing though high and which would undoubtedly lead to increased excess cancers in the population is rarely detected and so it is politically less sensitive. Essentially who cares as long as the patients don't realize and it doesn't end up in newspaper headlines?' Discuss the above statement including a consideration of the ethical issues involved.

1.13 Analytical, creative, practical and emotional intelligences: why you need all to succeed as a strategic and robust leader

The four types of intelligences relevant to Medical Physics leadership are analytical, creative, practical and emotional intelligences.

ANALYTICAL intelligence is the ability to analyse and evaluate ideas, solve problems and make decisions based on the evaluation of evidence. This is the type of intelligence which has traditionally been most revered (and with good reason) by Medical Physicists. It is crucial for Medical Physics work and for improving patient services. However, on its own *it is not sufficient to ensure successful leadership*!

CREATIVE intelligence involves the generation of new ideas, thinking outside the box and innovative ways of addressing issues, going beyond what is presently offered for discussion. This type of intelligence is not only important for the development of new medical devices or finding ground-breaking solutions for their use in solving clinical problems but also at finding creative ways of solving professional and educational issues.

PRACTICAL intelligence is the ability of leaders to find the best fit between their own strengths and weaknesses, the strengths and weaknesses of the group and the demands, opportunities and threats arising from the surrounding working, political and social environments. Leaders ignore such practicalities at their peril and put the group/team in danger. Many groups, teams and projects have foundered on leaders ignoring the importance of practical down-to-earth thinking.

EMOTIONAL intelligence is the ability to manage our emotions in a constructive intelligent manner so that they help us in life instead of making life more difficult for us. Emotions are important for us and are part of our humanity; but emotions can be both constructive and destructive. Even such a basic emotion as anger needs to be managed well. Quoting Aristotle: 'Anyone can become angry—that is easy. But to be angry with the right person, to the right degree, at the right time, for the right purpose, and in the right way—this is not easy'.

1.13.1 Reflection/discussion point

Try to analyse leaders that you have met in the past. Choose a range of successful, not so successful and bad leaders. Give them a score of 0–10 for the various types of intelligences. Score them also for their success as leaders. Check for correlation between the scores for the different types of intelligence and level of success.

Leader reference	Analytical intelligence score	Creative intelligence score	Practical intelligence score	Emotional intelligence score	Success as leader score

1.13.2 Reflection/discussion point

Discuss the following statement: 'Sometimes individuals try to expand the scope of their own role by copying or even poaching the role of others. This indicates a lack of creative and emotional intelligence'.

1.14 How can one prepare oneself for leadership roles?

Why do we need to prepare ourselves for leadership?—because society has become too complex for leading simply by intuition. Few of us are lucky enough to have been 'born leaders' (if such a thing actually exists) or had been given leadership development opportunities by family members or educators in our formative years. In addition, there is rarely (if any) systematic formal leadership training in most Medical Physics education and training programmes. Although many Medical Physics documents refer to the importance of leadership skills there are very few resources specifically targeted to Medical Physicists. If you are one of the majority—the unlucky ones who were not 'born leaders' or who were given few if any opportunities to develop leadership skills (welcome to the club!) it is not the end of the world. This book will help you become a strategic and robust leader. However, *please keep in mind that no book will turn you into a leader—ultimately YOU HAVE TO MAKE YOURSELF ONE.*

1.14.1 Reflection/discussion point

If you would like to become a good leader, you must first analyse yourself. Try to give an answer to the following questions:

- what personal characteristics do you consider that you have which will help you become a good leader?—these are your *strengths;*
- what personal characteristics do you consider could prevent you from becoming a good leader?—these are your *weaknesses*;
- set up a personal plan for developing your strengths further and eliminating your weaknesses or at least minimizing their effects on your leadership efforts.

1.14.2 Reflection/discussion point

If you would like to become a leader prepare a time-line for yourself. Try to give an initial answer to the following questions:

- what leadership role would you like to have in the short, medium, long-term?
- how can you prepare yourself to achieve your objective of becoming a strategic and robust leader in these roles? Take the wider view: consider education and training required, working with other leaders who can act as role models or mentors, membership of special interest groups, membership of national/regional/international professional committees/boards to gain experience, ethical considerations…

1.15 Leadership is a personal journey

Keep in mind that what will be actually achieved during your leadership will depend on:

- your own personal and professional history;
- your own personality, level of commitment, attitudes;
- the personality, level of commitment, attitudes of those you would be leading;
- the circumstances you find yourself in: history showers us with many leadership opportunities, but sometimes also unexpected challenges.

However, notwithstanding the challenges if you are willing to work hard, grasp opportunities, dive into it, you can achieve a lot.

1.16 Good leaders prepare future leaders

One of the trademarks of a good leader is the ability to *develop more leaders.* The young generation (and perhaps even the not so young) needs to be prepared to take over when it is time for the present generation of leaders to retire. If we do not prepare future leaders it will mean problems for the profession and hence patients in the future.

This can be done in many ways:

- being oneself an example of a good leader;
- role modelling—asking established leaders from within the department and from outside the department and even from other countries to present personal journeys, perspectives and experiences of leadership;
- encouraging younger members of the team to take on responsibility for specific tasks;
- preparing Medical Physics leaders by encouraging them to attend a formal leadership course specifically designed for Medical Physicists. Although there have been suggestions that joining an MBA programme would suffice [6], it is our opinion that this is too generic and programmes which are specifically tailored to the specific leadership learning and training needs of Medical Physicists would be more effective [7]. Read what professions closely related to ours are doing and learn from their experiences [8].

1.17 An essential fundamental to-do list for the strategic and robust leader

Let us end this with a handy to-do list for actual and potential leaders. The items in the list will be discussed more fully in later chapters but let us set your thinking going. Whichever type of group/team you are leading, whatever the environment here is an essential to-do list:

- clarify your vision (with the help of the group/team): what state would you like the group to be in by the end of your leadership tenure? Without having a clear well-thought-out vision as your constant guide and companion you will not be successful in improving the state of the group/team, instead the complexity of the healthcare system would deflect you from your purpose and perhaps wear down your enthusiasm.
- have a deep personal commitment: otherwise do the right thing—make way for someone else—Medical Physics is serious business—there is the patients' well-being at stake;
- conduct an honest self-analysis of personal strengths and weaknesses, ask yourself: 'do I need to change anything about myself to be successful in this leadership role—level of emotional intelligence, communication skills, attitudes, robustness?' Don't expect too much of yourself too quickly, one cannot become a leader overnight and changing oneself is a gradual process; be nice to yourself (but not too nice as changing oneself requires effort!);
- conduct an honest audit of the strengths and weaknesses of your group. Recognize their strengths but also their weaknesses as one cannot solve issues if one pretends they do not exist;
- decide on the style of leadership to adopt: if you are lucky with personalities/attitudes/motivation levels of group members happily democratic, if not benevolent authoritarian (what can you do?); in practice often a shifting blend of both;

- scan the environment for opportunities to help you achieve the vision and pay very careful attention to threats from the environment; ignoring threats would have disastrous consequences;
- work at it until the vision is achieved: we never achieve 100% of what we would like to do, but aim high, celebrate the ups, quickly accept the downs, amend your strategy if need be and move forward;
- have a succession plan: it is important that someone else continues the task when you leave, none of us is eternal, accept handing over to others kindly.

References

[1] Caruana C J, Cunha J A M and Orton C G 2017 Subjects such as strategic planning, extra-disciplinary communication, and management have become crucial to Medical Physics clinical practice and should become an integral part of the Medical Physics curriculum *Med. Phys.* **44** 3885–7

[2] Fontenla D P and Ezzell G A 2016 Medical physicist assistants are a bad idea *Med. Phys.* **43** 1–3

[3] Tamarat R and Benderitter M 2019 The medical follow-up of the radiological accident: Épinal 2006 *Radiat. Res.* **192** 251–7

[4] Peiffert D, Simon J M and Eschwege F 2007 Épinal radiotherapy accident: past, present, future *Cancer Radiother.* **11** 309–12

[5] Chustecka Z 2013 (February 6th) Docs in prison after radiation overdose in prostate cancer *Medscape Medical News*

[6] Gutierrez A N, Halvorsen Per H and Rong Y 2017 MBA degree is needed for leadership roles in Medical Physics profession *J. Appl. Clin. Med. Phys.* **18** 6–9

[7] Caruana C J *Leadership in Medical Physics, Development of the Profession and the Challenges for the MPE* www.eutempe.net.eu/mpe01 [accessed 5th January 2020]

[8] Davenport M S and Reed Dunnick N 2018 Lessons on leadership *RadioGraphics* **38** 1688–93

Chapter 2

A strategic planning primer for Medical Physics leaders

Learning outcomes

By the end of the chapter the reader will be able to:

- define strategic planning;
- explain in detail the meaning of the terms: mission statement, vision statement, strategic objectives, strategic plan;
- explain the critical role of vision in strategic leadership;
- distinguish between strategic planning, operational planning and project planning;
- explain the steps in developing, implementing and evaluating a strategic plan based on the SWOT methodology;
- understand that for any strategic planning exercise the identification of related SWOT issues should be the result of a research process which often involves multiple avenues involving both quantitative and qualitative research.

2.1 What is strategic planning?—in a single sentence please!

Readers with an interest in strategic planning and leadership are probably aware that there is much ambiguity in the definition of strategic planning in the literature so let us define it as succinctly as possible and with a minimum of ambiguity: *strategic planning is a systematic process for developing and achieving a desired future vision.*

doi:10.1088/978-0-7503-1395-7ch2

2.2 What is the difference between strategic planning, operational planning and project planning?

Planning can occur on three levels and leadership is required at all three. Strategic planning is a comprehensive exercise carried out say every five years. Operational planning is a yearly process based on the strategic plan in which the group expresses its objectives for that particular year. The yearly operational plan provides the members of the group with a road-map for the year and helps to guide their daily activities in relation to the more comprehensive strategic plan. A project plan provides details describing how a specific objective within the operational plan should be achieved. In this chapter we will focus on the process for developing a strategic plan. A method for the development of an operational plan can be found in [1].

2.3 Steps in developing a strategic plan

The steps in developing a strategic plan for a group/team vary in the literature. The sequence that we will be adopting in this book is given below. It is highly recommended that *all steps be carried out together with the members of the group/team* (or in the case of large groups elected representatives) as this would ensure that all will be aware of what is happening and the reasons why specific decisions were taken in a certain way. It also reduces problems arising from misunderstandings to a minimum. Very importantly, it will ensure that each member of the group/team feels a certain degree of ownership of the developed strategic plan. In addition, it gives space for any members who feel that they cannot for some reason or other contribute to the development of the vision and its implementation to state so and either take an acknowledged back-seat or leave the group/team. This would help avoid acrimonious feelings at a later stage, particularly in the implementation stage of the strategic plan.

Most frameworks for strategy development in common use today are based on some variant of the STRENGTHS–WEAKNESSES–OPPORTUNITIES–THREATS (SWOT) methodology. The sequence of steps for developing a strategic plan according to the SWOT methodology varies in the literature. The one we would be using is a simplified and slightly different version to the one used in [2]. The sequence is given below:

1. decide on the VALUES which would guide the strategic planning exercise;
2. develop a strategic MISSION STATEMENT for the group/team (or update the current one);
3. develop a desired FUTURE VISION for the group/team keeping the mission in mind;
4. carry out a GAP ANALYSIS, i.e. identify the gaps between the desired future vision and the current state of the group/team;
5. carry out a SITUATIONAL ANALYSIS in terms of the SWOTs of the group/team *with respect to the desired vision*;

6. identify and prioritize a set of STRATEGIC OBJECTIVES for closing the gaps, hence approaching the desired vision;
7. develop a detailed STRATEGIC ACTION PLAN to achieve each of the strategic objectives;
8. IMPLEMENT the plan;
9. EVALUATE and CONTROL the plan;
10. ITERATE.

We will now describe each step of the SWOT methodology in turn and discuss its application to specific medical physics situations

2.4 Decide on the VALUES which would guide the strategic planning exercise

Underpinning your strategic planning are your own personal values and the values of the group/team that you would be leading. Let your values guide you and keep you moving along a trustworthy and ethical path. Here is a possible set of values—add, amend and reword according to your own values and preferences:

1. Search for EXCELLENCE: we consistently pursue and maintain the highest quality of clinical service, education and research;
2. RESPECT: we respect the dignity, uniqueness and particular health, educational and professional needs of every patient, student, trainee and healthcare profession;
3. PROFESSIONALISM: we work with expertise, commitment, integrity, fairness, reliability and flexibility, hence providing role modelling opportunities for younger members of staff;
4. SERVICE: we maintain positive relationships when fulfilling our duties with respect to the expectations of patients, students and other healthcare professionals who request our services;
5. TEAMWORK: we recognize that quality healthcare and professional education are intrinsically inter-professional activities;
6. LIFELONG LEARNING: we affirm that in a knowledge-based society, excellence, professionalism and service are based on a process of lifelong learning;
7. PATIENT ORIENTATION: we keep in mind that the ultimate focus of all our activities is the individual patient.

2.5 Develop a strategic MISSION statement for the group (or update the current one)

The mission of a group is basically a statement which answers the following questions:

- *who* benefits from the activities of the group?
- *what unique* services does the group provide to its clients?

- in the case of groups which form part of a larger association/organization (e.g. medical physics department of a hospital or faculty): how does the group contribute to the over-arching mission of the wider organization of which it forms part?

A good mission statement should have as many of the following characteristics as possible:

- includes those competences[1] which are *uniquely or best provided by the group above all others* (particularly any competitors). In the strategic planning literature these are known as the 'CORE COMPETENCES' of the group;
- mentions the principal clients of that profession. In the case of medical physics in general our main clients are patients (including asymptotic individuals, carers and comforters, volunteers taking part in medical research, and others imaged using medical radiological equipment (EU Dir 2013/59/EURATOM Articles 2(48) and 22(4)) [3]. However, there are other possibilities e.g. hospital management or other healthcare professions as we provide advice and educational services to them (EU Dir 2013/59/EURATOM Article 83(2)(h)) [3];
- makes use of the same terminology used in legal documentation as it adds weight to the mission statement, however, avoid very difficult or abstract terminology so as to make the mission statement understandable by as large a number of clients or potential clients as possible;
- is comprehensive enough to demonstrate the entire range of core competences of the group/team;
- is forward looking and gives pride and direction to group/team members including direction regarding further development of the mission;
- not be overly long so that it is easier to remember and place on homepages, posters, stationery, entrances to departments, banners …

It is not always possible for all the above characteristics to be satisfied simultaneously—for example, sometimes it is difficult to be both comprehensive and yet concise!

A good mission statement is a strong marketing tool—it tells clients 'this is what we are doing for you!' It should be on the homepage of members of the department/organization, departmental stationery, entrance to the department, email signature of group members etc so that as many people as possible get to know why the activities of the group are important. *It is no use having the best and most relevant mission in the world if nobody knows about it!*

[1] Unfortunately, the word 'competence' is used with different meanings in the literature. In this book the word 'competence' is used with the meaning of 'responsibility and autonomy' as defined in the European Qualifications Framework.

2.5.1 Reflection/discussion point

Try to design a mission statement for each of the following:

- the Medical Physics profession at the global level (as you think it should be formulated by the IOMP);
- the Medical Physics profession at the level of your region (as you think it should be formulated by say EFOMP, AFOMP);
- the Medical Physics profession at your national level (as you think it should be formulated by your national Medical Physics organization);
- the Medical Physics department team at your hospital (as you think it should be formulated by the members of your Medical Physics department).

Now compare with any actual mission statements developed by the respective organizations/groups (if any) as indicated in the reflection/discussion points below.

2.5.2 Reflection/discussion point

The international (global) mission statement for the Medical Physics *profession* provided by the IOMP is: 'Medical Physicists apply physics in medicine'[2]. Evaluate it in terms of the above characteristics. Evaluate it in particularly as a marketing tool for the profession. Put yourself in the shoes of officials from ministries of health, ordinary people in the street (present/future patients) and other healthcare professions. Do you think it offers sufficient insight into what we do as a profession to these classes of stakeholders and in a language they can understand?

2.5.3 Reflection/discussion point

Read the following mission statement for the Medical Physics *profession in Europe* provided by EFOMP. Evaluate it in terms of the characteristics of a good mission statement including the appropriateness of language *vis à vis* relevant stakeholders.

'Medical Physicists and Medical Physics Experts will contribute to the maintaining and improving of the quality, safety and cost-effectiveness of healthcare services through patient-oriented activities requiring expert action, involvement or advice regarding the specification, selection, acceptance testing, commissioning, quality assurance/control and optimized clinical use of medical devices and regarding patient risks from associated physical agents including protection from such physical agents, installation design and surveillance, and the prevention of unintended or accidental exposures to physical agents; all activities will be based on current best evidence or own scientific research when the available evidence is not sufficient. The scope includes risks to volunteers in biomedical research and carers and comforters' [4].

Note that this mission statement expands the mission (scope of the role) to all medical devices (not just radiological devices) and to all physical agents (not just ionizing radiation).

[2] https://www.iomp.org/wp-content/uploads/2019/02/iomp_policy_statement_no_1_0.pdf.

2.5.4 Reflection/discussion point

Consider the following potential very concise global mission statement for the global medical physics *profession* based on a suggestion by a participant of a past EUTEMPE leadership course; evaluate it in terms of the characteristics of a good mission statement.

'Medical Physicists ensure accuracy, precision and safety in the clinical use of medical devices and physical agents in healthcare'.

2.5.5 Reflection/discussion point

The mission of a Medical Physics organization *per se*[3] is to lead Medical Physicists who are members of that organization (either directly or through representatives). The present mission statement of the IOMP is the following:

'The mission of IOMP is to advance Medical Physics practice worldwide by disseminating scientific and technical information, fostering the educational and professional development of Medical Physicists, and promoting the highest quality medical services for patients'[4].

Consider the following modified mission statement for the IOMP and discuss its advantages and disadvantages with respect to the present mission statement.

'The mission of IOMP is to help advance quality healthcare services for all patients worldwide by fostering the professional development of Medical Physicists globally through education, research and partnerships with international stakeholders'.

2.5.6 Reflection/discussion point

The mission of a Medical Physics association/organization *per se* is to lead Medical Physicists who are members of that organization (either directly or through representatives). The mission statement of the AAPM is the following:

'The mission of the Association is to advance the practice of physics in medicine and biology by encouraging innovative research and development, disseminating scientific and technical information, fostering the education and professional development of Medical Physicists, and promoting the highest quality medical services for patients'[5].

Can you think of ways of improving it?

[3] It is important to distinguish between the mission of an organization *per se* and the mission of the professionals which it represents. The mission of IOMP as an organization *per se* is the well-being of *Medical Physicists*, the mission of Medical Physicists is the well-being of *patients*.

[4] https://www.iomp.org/organisation/.

[5] https://www.aapm.org/org/objectives_ev.asp.

2.6 Develop a desired future VISION for the group keeping the mission in mind

A vision statement is the answer to the following question: 'Whilst keeping the mission in mind, what state would the group/team like to be in five/ten/fifteen years' time?' (short/medium/long term visions).

The vision statement is a statement describing in a brief manner the desired future state for the group/team to aspire to. The vision statement is very important for you as a leader as it will be your constant guide in your everyday leadership challenges and issues. You will use it to assess the state of the group/team that you are leading and to set your objectives. Your vision is your guide through the good times and the bad, the ups and downs. It keeps you from losing your way in the world. The vision should be developed in collaboration with the group/team so that ownership of the vision is shared by all.

A good vision statement should if at all possible include the following characteristics as a minimum:

- it should follow from the mission statement to ensure relevancy to clients and stakeholders;
- be future oriented;
- be inspiring and motivating for the group/team members;
- be perceived to be achievable by the group/team members.

One way of creating a vision statement is to write an article that you would love to see published about your group in say ten years from the present. From the article, pull out three to five vivid snippets that you feel bring your envisioned future vision to life. Use these snippets to create the vision statement.

2.6.1 Reflection/discussion point

Here is a mission statement for a team of biomedical physics educators in the teaching of healthcare professionals. Evaluate it in terms of the characteristics of a good mission statement. Can you improve it in any way? Note that the term 'biomedical physics' is used in this context instead of Medical Physics as the scope of practice of this particular group would not only include the knowledge and skills traditionally associated with Medical Physics (i.e. use of radiation in imaging and therapy) but also such topics as microscopy, physiological measurement, biophysics and others.

'We will make a key contribution to quality healthcare professional education through knowledge transfer activities concerning the techno-scientific knowledge, skills and competences supporting the clinically-effective, evidence-based and economical use of biomedical devices and safety issues concerning associated physical agents. Our efforts will be guided by an appreciation of the value of all healthcare professions and underpinned by research-based curriculum development' [2].

Now here is a vision statement for the team of biomedical physics educators. Evaluate it in terms of the characteristics of a good vision statement (including whether it is sufficiently linked to the previous mission). How can it be improved?

'The biomedical physics educator will be recognized by the educational leaders of all healthcare professions across Europe as the educator of first call with respect to the techno-scientific knowledge/skills/competences underpinning the clinically-effective, evidence-based and economical use of biomedical devices and safety issues concerning associated physical agents and be perceived as providing a practice-oriented, learning-outcomes based, well-integrated, research-based, internationally-harmonized, ethically and inter-professionally oriented contribution to the education of healthcare professional students' [2].

2.6.2 Reflection/discussion point

Many people unfortunately confuse mission and vision statements and some even use the two terms interchangeably—yet they are not the same and the purpose of each is different. Can you distinguish clearly between them at this stage? Discuss the statement 'vision follows mission'.

2.7 Carry out a GAP ANALYSIS

A gap analysis refers to the identification of the differences between the desired FUTURE state of the group/team as enshrined in the vision statement and the actual PRESENT situation of the group/team. The aim of the strategic plan will be to eliminate these gaps and hence achieve vision. It is essential to be honest and make sure that all gaps are identified as gaps which are not will simply not be addressed leading to an eventual underachieving of vision. The wording of the gaps should be derived directly from the words used in the vision statement so that vision and gaps are in synchrony.

2.7.1 Reflection/discussion point

Consider the study on the role of the biomedical physics educator described in [2, 5, 6]. Read carefully the mission, vision and gaps. *Analyse how the wording of the gaps was derived directly from the words used in the vision statement.*

2.8 Carry out a SWOT SITUATIONAL ANALYSIS for the group with respect to the desired vision

SWOT (STRENGTHS–WEAKNESSES–OPPORTUNITIES–THREATS[6]) refers to those issues which would need to be considered in order for the strategic plan to be developed:

---

[6] Some authors prefer the word 'challenge' than 'threat' as it indicates something to overcome instead of the more daunting word 'threat' and even use SWOC instead of SWOT. We would also prefer using the word 'challenge', however the acronym SWOT is the one still used mostly in the literature and we will stick to this to avoid confusion.

- STRENGTHS: what are the strengths that your group/team has that will help you eliminate gaps and achieve vision. Very important strengths are your CORE COMPETENCES, i.e. those competences and associated knowledge and skills in which you have a clear advantage over all competitors;
- WEAKNESSES: what are the weaknesses that your group/team has that will hinder you from achieving your vision? Very important are the CRITICAL WEAKNESSES[7]. Critical weaknesses are the opposite of core competences—whilst core competences ensure that a group thrives, critical weaknesses are a disease that can destroy it from within unless treated. You must analyze these very carefully because if not tackled they would undermine your success. Psychologically, it is easier to think of our strengths and we dislike thinking of our weaknesses—however, this is very dangerous as weaknesses if not addressed and eliminated can destroy your capability of achieving the vision;
- OPPORTUNITIES are positive issues from the surrounding Political, Environmental, Socio-psychological, Techno-scientific, LEgal (PESTLE) milieus that you can capitalize on in order to achieve your vision. Which are your KEY OPPORTUNITIES, i.e. those opportunities with the higher benefit value?
- THREATS: what are those negative factors from the surrounding Political, Environmental, Socio-psychological, Techno-scientific, LEgal (PESTLE) milieus that if not tackled would hinder you from achieving your vision? Which are the EXISTENTIAL THREATS—those threats the consequences of which are so severe that they can destroy your vision completely? *These must be eliminated.*

2.8.1 Reflection/discussion point

Identify SWOT themes for each of the following using the generic SWOT matrix shown in table 2.1:

- the Medical Physics professional group in your country;
- the departmental team at your hospital;
- the Medical Physics professional group in your region;
- the Medical Physics professional group globally.

In the next four chapters you will find a compendium of SWOT themes relevant to the Medical Physics profession to help you fill the SWOT matrix, but at this stage try to come up with ideas yourself without consulting the compendium.

2.8.2 Reflection/discussion point

Develop your own personal SWOT matrix keeping in mind your desired vision for yourself (i.e. where would you like your professional career to be in say five/ten years' time?).

[7] Some authors call these 'core problems'.

Table 2.1. A basic SWOT matrix.

	Positive factors	Negative factors
STRENGTHS and WEAKNESSES of your group/team	What are the strengths of your group/team? What are the core competences? What are the strengths with respect to competitors? Is there a difference between the self-perceived strengths of your group/team and the way other stakeholders see them?	What are the weaknesses of your group/team? What are the critical weaknesses? What are the weaknesses with respect to competitors? Is there a difference between the self-perceived weaknesses of your group/team and the way other stakeholders see them?
OPPORTUNITIES and THREATS for your group/team from the surrounding environment (use the PESTLE acronym to help)	What current opportunities emanate from the environment? What future opportunities will be available in practice? What are the key opportunities? What opportunities can your group/team create for itself?	What current threats emanate from the environment? What possible future threats do you think will come up? What are the existential threats? Is your group/team facing unfair competition?

2.9 Identify and prioritize a set of STRATEGIC OBJECTIVES for closing the gaps, hence approaching the desired vision

The strategic objectives are those objectives that would lead to the achievement of vision; they would be statements of the type:

- to reinforce this core competence we will…
- to eliminate this weakness we will…
- to benefit from this opportunity we will…
- to eliminate or reduce the impact of this threat to below the vulnerability level we will….

Your list of strategic objectives should include ways the group/team can make the most of existing or expected future opportunities (and whenever possible the creation of new opportunities by the group/team itself) for the further strengthening of strengths and the minimization or complete elimination of weakness and threats *with particular attention being given to core competences, critical weaknesses and existential threats.*

It is important to distinguish between *critical strategic objectives and desirable but not essential ones and prioritise the former.* Make sure that the list of strategic objectives can be achieved in the time available—*having too many objectives is one way of ensuring that many would not be achieved with the attendant risk of the failure of the plan as a whole!*

2.10 Develop a detailed STRATEGIC ACTION PLAN to achieve each of the strategic objectives

A strategic action plan is a documented plan for achieving each of the strategic objectives including processes, assignment of resources and costings. Each objective would need to be addressed and a way must be found to ensure that there is a high probability that the objective would be achieved. Here are some essentials to reflect upon:

1. plan structure: have you prioritized objectives and ensured that objectives will be addressed in a logical sequence?
2. is the plan integrated in the sense that all elements of the plan for all objectives support each other and that there are no conflicting elements?
3. do you have a time-line for the plan which includes target dates for each objective?
4. is the plan realistic given the available time? Consider pushing some deadlines out further;
5. is it too complex and rigid? Overplanning is a common problem;
6. have you considered different possible (what if?) scenarios? Is there a plan B for the most critical objectives?
7. have you included the material, human and financial resources necessary?
8. have you included any further training that either the group as a whole or individual members of the group may need to be able to support the plan?
9. have you included Key Performance Indicators (KPIs)? These are defined as metrics/criteria which would indicate the level of progress in the implementation of the strategic plan (make sure you only include a reasonable number of these, you don't want to spend half your working (or perhaps waking!) hours monitoring KPIs instead of actually getting on with the job!);
10. is the plan complete? Identify any lacunae in your plan or potential activities that are not supported;
11. is the document clear and easily understood? Are there parts which are ambiguous and open to excessive individual subjective interpretation?

2.10.1 Reflection/discussion point

References [2, 5, 6] describe an EFOMP project for the systematic development of a research-based strategic plan for the role of the biomedical physics educator in the education of healthcare professionals (physicians, nurses and others) in Europe. Study the articles properly in depth:

- why was it important to start the strategic planning with a literature review?
- how did the authors use research to inform their SWOT analysis?
- can you identify each step of the development of the strategic plan?

2.11 IMPLEMENT the plan

The implementation of the plan must be guided at all times by relevant questions such as:

- to achieve our vision what must we excel at?
- to achieve our vision how should we appear to our stakeholders?
- to achieve our vision how will we sustain our ability to change and improve?
- to achieve our vision what resources do we need?

The implementation stage is often the most difficult stage of strategic management. Here are some hints to make it happen:

1. assign specific objectives to specific task groups (actually the term 'task team' would be more appropriate in this context). Ensure that tasks are well matched to the knowledge and skills of the particular task group members;
2. ensure that formal and informal communication lines are set up and maintained within and across the task groups: regular (frequency of meetings depending on the type of group), structured discussions regarding objectives and progress of implementation involving parts of or the entire group should be the norm;
3. align the contributions of the different members of the group and various teams: every member of the group should understand the current state of things, the vision destination, and the journey. Everyone should appreciate the relevance of own contribution to that journey and appreciate and acknowledge the part played by others;
4. drive motivation of the group members: without sufficient motivation nothing gets done. Leaders should never underestimate the importance of the motivational level of a group for achieving objectives;
5. maintain focus: do not get sidelined by trivialities or secondary issues which sap your strength;
6. be action-oriented: a common weakness in executing organizational, department and personal goals is not taking note of daily actions to reach the desired goal. Everyone's daily and weekly focus should be on the very specific tasks they need to achieve to move their goals forward, and those goals need to align with the overall vision;
7. develop educational activities and policies to support implementation. At each stage consider whether some members of the group require education/training in specific areas in order to be able to make a solid contribution;
8. address lack of resources—members of the group need to devote their time to the important tasks at hand, not waste their mental energies searching for funds;
9. think about how to reduce resistance to change which is always a significant threat, particularly in large groups.

The 'balanced scorecard' frameworks used for tracking and managing organizational strategies in the business world can be easily modified for strategic plan implementation, tracking and management by the professions. By 'balanced' one means that there is a good balance between stakeholder demands, internal processes and financial and learning issues and that one does not focus on say the financial to the detriment of all other factors [7].

2.12 EVALUATE and CONTROL the plan

Evaluation and control is the process of determining the effectiveness of a given strategic action plan in achieving the strategic objectives. This should be carried out on an ongoing basis and corrective actions taken whenever required. It is important to formulate a way of finding out and monitoring whether the action plan being implemented is effective and efficient[8] in achieving the strategic objectives. Objective criteria/KPI metrics for assessing effectiveness and if possible efficiency should be monitored. The criteria/KPI should be linked directly to the strategic objectives. On the other hand, keep it simple. Do not overload the group members with too many criteria/KPIs to monitor. Strategic KPIs should be used at 'board-room' level to assess the achievement of the strategic objectives, that is, monitoring where the group is at a given point in time in relation to the future vision. These are different from operational ('shop-floor' level) KPIs that are used to assess normal operational delivery on a daily basis. It goes without saying strategic and operational KPIs should be aligned.

2.13 ITERATE

In the rapidly changing world we live in, strategic planning is an ongoing process. Strategic plans need to be updated and modified continuously to address changes in the external environment which may impact the SWOTs and hence the vision and its attainment.

2.13.1 Reflection/discussion point

Articles [2, 5, 6] are now a bit dated. Do you think that the strategic objectives are still valid as they stand or need to be modified for changes that have occurred since then? Criticise the articles constructively—how can the strategic plan developed in the articles be further improved?

2.14 A SWOT compendium resource for Medical Physics strategic planning

The SWOTs for the group/team with respect to the vision should be the result of a research exercise carried out by the leadership together with the members of the group/team. The methodologies used should involve not only quantitative methodologies (e.g. surveys) but also very importantly qualitative ones (e.g. interviews,

[8] We distinguish between effectiveness and efficiency. The former is the extent to which the strategic objectives are being achieved or otherwise, the latter is the extent to which the objectives are achieved with the minimum use of resources and the avoidance of waste.

focus groups, documentary analysis, nominal group technique, Delphi technique). Often mixed methods of research involving both qualitative and quantitative methods would be necessary to capture all the relevant SWOT issues. Use of such mixed approaches is increasing in Medical Physics as research in educational and professional issues expands. Research participants should include not only members of the group itself but also relevant external stakeholders. No group exists in isolation and the perspectives of such stakeholders need to be identified and taken into account if positive, learned from if relevant and countered if negative. Although such methods are time consuming, in the long run one would be confident that the strategic planning exercise would be built on solid foundations and an accurate and realistic assessment of the situation.

The next four chapters will present a compendium of generic SWOT themes for the Medical Physics professional group as a whole. This will provide leaders with a ready resource of possible ideas to trigger the initial SWOT brainstorming and discussions with their specific group/team. *The use of the compendium should reduce markedly the time required to complete the SWOT audit.* The compendium is based on the available literature (quite sparse) and the many discussions by the author with Medical Physics colleagues over the past 30 years in national, European and international fora. *From this compendium the group should choose the SWOTs which are relevant to their group and use this initial set to develop a more comprehensive audit which is specific to their own situation.* The SWOT issues have been categorised for the convenience of the reader in the usual strengths–weaknesses–opportunities–threats categories. We will discuss each of these SWOT themes in turn. A word of caution: note that a specific strength of the Medical Physics profession as a whole can be weak in a given particular subgroup e.g. a small Medical Physics group in a small hospital can be lacking in expertise of acceptance testing procedures of CT or MRI scanners or statistical skills, or the profession may not yet be legally recognized in a given country. *A strategic plan is devised for a particular group/team and must take local variabilities into account in order to be successful.*

References

[1] Towbin A J, Perry L A, Moskovitz J A and O'Connor T J 2018 Building and implementing an operational plan *RadioGraphics* **38** 1694–704

[2] Caruana C J, Wasilewska-Radwanska M, Aurengo A, Dendy P P, Karenauskaite V and Malisa M R *et al* 2012 A strategic development model for the role of the bioMedical Physicist in the education of healthcare professionals in Europe *Phys. Med.—Eur. J.Med. Phys.* **28** 307–18

[3] European Council 2013 *Council Directive of 5 December 2013 Laying Down Basic Safety Standards for Protection Against The Dangers Arising from Exposure to Ionising Radiation, and Repealing Directives 89/618/Euratom, 90/641/Euratom, 96/29/Euratom, 97/43/Euratom and 2003/122/Euratom, Official Journal of the European Union L-13 of 17/01/2014.*

[4] Caruana C J, Christofides S and Hartmann G H 2014 EFOMP Policy Statement 12.1: recommendations on medical physics education and training in Europe 2014 *Phys. Med.—Eur. J. Med. Phys.* **30** 598–603

[5] Caruana C J, Wasilewska-Radwanska M, Aurengo A, Dendy P P, Karenauskaite V and Malisa M R *et al* 2009 The role of the bioMedical Physicist in the education of the healthcare professions: An EFOMP project *Phys. Med.—Eur. J. Med. Phys.* **25** 133–40

[6] Caruana C J, Wasilewska-Radwanska M, Aurengo A, Dendy P P, Karenauskaite V and Malisa M R *et al* 2010 A comprehensive SWOT audit of the role of the biomedical physicist in the education of healthcare professionals in Europe *Phys. Med.—Eur. J. Med. Phys.* **26** 98–110

[7] Wikipedia Balanced Scorecard Framework https://en.wikipedia.org/wiki/Balanced_scorecard [accessed 20 December 2019]

Chapter 3

Internal STRENGTHS of Medical Physics

Learning outcomes

By the end of the chapter the reader will be able to:

- discuss the main strengths of Medical Physics and their importance for strategic planning in Medical Physics;
- understand more deeply the meaning of the term 'core competence';
- discuss the core competences of the Medical Physics profession;
- understand why it is crucial that core competences be strengthened.

3.1 Strengths, core competences and grasping opportunities

The strengths of a profession are those intrinsic characteristics of the profession that will be conducive in helping it achieve its vision. The CORE COMPETENCES of the profession are those capabilities that the profession is very strong in, in fact so strong that competitors would find it extremely hard or impossible to acquire. Most of the strengths of the profession discussed in this chapter are core competences. Core competences should be fostered and developed to a point of perfection as they are what makes a profession what it is and serve to protect it in a world of competition, fake assertions, austerity economics and unbridled commoditization. A strategic plan must include actions for making the most of opportunities to strengthen core competences if they are weak among the members of the group/team or strengthen them even further to a level totally unachievable by competitors.

3.2 Core competence: deep techno-scientific expertise regarding medical devices and their clinical use

No other profession in healthcare has the deep techno-scientific expertise regarding medical devices and their clinical use than Medical Physics (and its sister profession

doi:10.1088/978-0-7503-1395-7ch3

Biomedical Engineering when the latter has a patient-centred clinical orientation as opposed to an orientation towards industry). Certainly in the case of the use of ionising and non-ionising radiations in medicine most techniques have been developed and are still being developed by Medical Physicists. It is indeed no coincidence that the foremost clinical D&IR, NM and RO departments in the world are associated with the presence of a strong Medical Physics department and that the weakest departments in the world are such because they lack one. Many legislators in the world have realized that the ever increasing adoption and use of complex physics based medical device technologies in healthcare today requires the oversight of highly techno-scientific professionals with a sufficiently deep understanding of the underlying physics and functioning of the devices involved. It is an acknowledged fact that these technologies are developing at a much faster rate than the education and training of the non-physics healthcare professionals using them and that the same healthcare professionals are stressed because they know that their expertise is insufficient for effective and safe practice [1]. Frequent, ongoing, expensive retraining is found to be necessary. Unfortunately, this retraining (often by non-local 'applications specialists' with little commitment to the local milieu) is rarely as effective as it should be, as the healthcare professionals lack the necessary physics and mathematical knowledge. It is again often difficult for healthcare managers to release these professionals to attend continuous professional development classes (CPD) owing to the long waiting times for patients to be serviced. In such circumstances the most efficient way forward in terms of both time and expense is to have onsite Medical Physicists who are capable of doing the necessary study of user and service manuals for newly introduced devices *under their own steam* and without excessive reliance on the ever more expensive 'application specialists' whose training interventions are in practice relatively short and minimal. Owing to the lack of real understanding of the physics, the use of medical devices by healthcare professionals is rarely as effective and safe as it should be and certainly not efficient —with devices often being used at a small fraction of their capabilities. This is particularly true of the more complex devices such as CT, MRI and linacs. Many potential uses of the devices remain unexplored leading to underdiagnoses and insufficiently effective radiation therapy. Often these professionals, either due to excessive professional pride or stubbornness, do not ask for advice, some of them do not even realize they need advice [2]. There is a waste of expensive resources and money lost in litigation leading to a marked effect on healthcare budgets. In addition, too much money is wasted by hospitals on external consultancies involving the procurement/introduction of new devices in the clinics and the oversight of software updates which could be done by Medical Physicists—and not simply on an intermittent basis by external consultants with sometimes little commitment to the healthcare of the local population.

Millions of euros are spent annually on research into the clinically-effective, safe, efficient and optimised use of medical devices and thousands of papers published— yet so few of the results are translated into actual clinical service development as healthcare professionals are incapable of understanding the research articles (and what is even worse, may even misunderstand them). The results from these studies would save thousands of lives, improve healthcare safety and save millions in unnecessary clinical procedures or procedures which would have been avoided if the disease had been detected earlier (e.g. the high number of cancer cases resulting from the late detection of cancers owing to insufficiently quality controlled imaging equipment). A major part of the problem is that the percentage of healthcare professionals who are capable of understanding the research literature and incorporating the results in their clinical work leaves, to put it mildly, much to be desired. Owing to the low uptake of the hard sciences by students (even at the pre-tertiary level), the situation is reaching critically epic proportions. In many countries healthcare professionals using sophisticated devices which are difficult even for physicists to understand do not even have an understanding of the basic physics principles involved.

3.3 Core competence: wide-ranging expertise concerning the protection of patients, workers and the general public from ionizing radiation and other physical agents in healthcare

Physical agents (which is the legal term used when referring to sources of energy such as ionising and non-ionising radiations, particle beams, ultrasound, vibration, optical sources, laser, heat sources) are termed so because their scientific study is in the domain of physical science[1] [3]. These agents are of immense benefit in medicine and their use is constantly on the increase. On the other hand, however, there is also an element of risk and when used by people with insufficient knowledge of their biophysical properties they become also a source of unacceptable or unnecessary risk. Medical Physicists have a long history of involvement (and often in leadership roles) at the national, European and international level in the development of protection guidelines with respect to such agents. The example that comes to mind is of course that of ionising radiation where Medical Physics professionals have been by far the most heavily involved group. Expert task groups within the International Commission on Radiological Protection (ICRP)[2] and the International Commission on Non-Ionizing Radiation Protection (ICNIRP)[3] are

[1] The EU has promulgated several directives in its 'physical agents' series of directives regarding occupational safety [3]. Interestingly, the ionizing radiation directive 2013/59/EURATOM which relates to the most carcinogenic of all physical agents is not positioned under this series. The reason is administrative with shades of the political—whilst the physical agents directives are under the auspices of Directorate-General for Employment, Social Affairs and Inclusion of the European Commission, the ionizing radiation directive which addresses all uses of ionizing radiation from medical to energy is owing to its association with nuclear power under the auspices of Directorate-General for Energy. The physical agents directives are all occupational health based whilst the ionizing radiation directive addresses occupational, public and patient safety.

[2] www.icrp.org.

[3] https://www.icnirp.org/.

often led by physicists and physicists are unfailingly involved when incidents occur both in the medical and non-medical use of such agents [4].

3.4 Core competence: strong analytical, problem-solving and trouble-shooting skills

Clinical departments at all levels but particularly at tertiary and quaternary levels require high level professionals with strong analytical, problem-solving and trouble-shooting skills. Few (if any) university programmes involve the learning of such skills to the level of a degree in physics or engineering. Indeed, only students with a high level of the aforementioned skills survive these demanding programmes—in essence if you do not have intrinsic high cognitive analytical abilities, high motivational levels and a readiness to work hard you are as good as dead. Graduates from these programmes will be found in all areas of the economy where analytical thinking is a pre-requisite for effectiveness and success—from hard science laboratories to financial institutions and the boards of large companies. Most other healthcare professions have only relatively recently recognized the importance of giving their students practice in problem-solving skills whilst in physics and engineering education problem-solving has always been an intrinsic part of the education and training. In fact, every chapter of every physics or engineering textbook ends with a list of problems to solve, all of which require analysis and a deep level of thinking. This is one of the reasons why a Bachelor's degree with a very high physics component (and associated mathematical content) is a requirement for certification as a Medical Physicist [5–7]. The rationale is basically ethical—Medical Physics work affects the health of millions of people worldwide; it is too important to be left in the hands of those who do not have the necessary analytical skills. This is not an attempt at the glorification of the profession—it is simple down-to-earth practicality. Those of us who have been involved in the profession for many years are aware of the issues arising wherever a department lacks the services of a good Medical Physics department. We all have experienced images of degraded quality which nobody seems to detect and the overdosing or underdosing of radiotherapy patients. Incidents and accidents in radiotherapy are often detected simply owing to the acute easily visible deterministic effects. However, incidents and accidents in imaging are rarely detected, as the detrimental effect on the individual patient is not acute and therefore not immediately discernible—the result is millions of overdosed patients. Many of the reported radiation incidents are really failures resulting from insufficient analysis and problem-solving skills. The requirement for Medical Physicists to have a first degree with a high physics and mathematics content is not simply the need for a wide base of physics and mathematics knowledge and skills for effective and safe Medical Physics professional practice but also to act as a filter to ensure that only those of sufficiently high calibre make it into the profession.

3.5 Core competence: strong mathematical, statistical and data analysis skills

No other healthcare profession approaches even remotely Medical Physics when it comes to mathematical, statistic and quantitative data analysis skills. Indeed, our skills in these areas are very much sought after by members of the healthcare professions. This is not only true in those clinics which include a research programme but even in non-research oriented clinics. The increased clinical applications of quantitative imaging, artificial intelligence and machine learning is leading to another higher level dimension [8]. Indeed, physics is inherently quantitative and a physicist without deep application oriented mathematical skills is a non-starter. Artificial intelligence and machine learning is being included in Medical Physics curricula[4]. Those of us who have been teaching healthcare professionals are aware of the problems we face when teaching such professionals owing to their low mathematical skills (often verging on the phobic). Often we have to divest our lectures from almost every equation including the simplest. Problems arise when healthcare professionals use equations and statistical tests without real understanding. Many reviewers of healthcare professional journals themselves are incapable of assessing the validity or otherwise of the equations and statistical techniques used with the result that quantitative mistakes are ubiquitous in the research literature of the healthcare professions.

3.6 Solid legal foundations for the profession

The Medical Physics profession has a strong legal foundation in many countries of the world. This is particularly true of the use of ionising radiation in medicine. The IAEA Basic Safety Standards emphasizes the need for Medical Physics input and the IAEA itself has produced many publications for the education and training of Medical Physicists [9–14]. At the European level the EU has made the presence of Medical Physicists at the expert level (Medical Physics Expert, MPE)[5] mandatory in hospitals and clinics. The legal provisions on the role of the Medical Physicist in

[4] At the moment (2019) a curriculum in artificial intelligence and machine learning for Medical Physicists is being developed by a SIG of the EFOMP.

[5] European legislation distinguishes two levels of Medical Physics expertise: an entry-to-the-profession level (Medical Physicist) and a high expert level (MPE). In order to provide a higher level of protection, EU legislation focusses on the second higher level. In addition, the directive distinguishes between MPE and radiation protection expert; the former relates to exposure of patients (including asymptotic individuals, carers and comforters, volunteers taking part in medical research, and others imaged using medical radiological equipment), whilst the latter relates to occupational and public exposure. An MPE can also be a radiation protection expert if he/she is so qualified, indeed in small hospitals this is often the case.

Europe in the area of ionizing radiation can be found in article 83[6] of European directive 2013/59/EURATOM. Further elaboration of the role of the MPE based on the EU directive can be found in the 'European Guidelines on the Medical Physics Expert' document [15]. An update of both documents can be found in EFOMP Policy Statement 16 [2]. The fact that the profession has a strong legal foundation is a major strength for the profession in Europe and makes it possible to access European funds to help develop the profession. Indeed, the EFOMP has regularly succeeded in tapping such funds.

3.7 Strong scientific research skills and highly qualified academics

Medical Physicists are in possession of unparalleled scientific quantitative research skills honed over many years of education. For us issues of accuracy, trueness, precision, reliability, estimates of uncertainty (alternatively known in the statistical literature as 'confidence intervals') are not concepts discussed during a few hours of lectures in a study unit called 'Research Methods' in the final year or pre-final academic year of an undergraduate course as is the case of most healthcare professionals. We live and breathe these concepts daily over the many years of our long education and training. It is little wonder that the most scientifically rigorous articles in imaging or therapy are found in physics or engineering journals or penned by physicists or engineers.

[6] Article 83 of EU Directive 2013/59/EURATOM.

Medical physics expert

1. Member States shall require the medical physics expert to act or give specialist advice, as appropriate, on matters relating to radiation physics for implementing the requirements set out in chapter VII and in point (c) of Article 22(4) of this Directive.
2. Member States shall ensure that depending on the medical radiological practice, the medical physics expert takes responsibility for dosimetry, including physical measurements for evaluation of the dose delivered to the patient and other individuals subject to medical exposure, give advice on medical radiological equipment, and contribute in particular to the following:
   - (a) optimisation of the radiation protection of patients and other individuals subject to medical exposure, including the application and use of diagnostic reference levels;
   - (b) the definition and performance of quality assurance of the medical radiological equipment;
   - (c) acceptance testing of medical radiological equipment;
   - (d) the preparation of technical specifications for medical radiological equipment and installation design;
   - (e) the surveillance of the medical radiological installations;
   - (f) the analysis of events involving, or potentially involving, accidental or unintended medical exposures;
   - (g) the selection of equipment required to perform radiation protection measurements;
   - (h) the training of practitioners and other staff in relevant aspects of radiation protection.
3. The medical physics expert shall, where appropriate, liaise with the radiation protection expert.

Note that in Europe we distinguish two levels of Medical Physics expertise: an entry to the profession level (Medical Physicist) and a high expert level (Medical Physics Expert). The EU Directive focusses on the second level. In addition, the directive distinguishes between Medical Physics Expert and Radiation Protection Expert; the former relates to exposure of (including asymptotic individuals, carers and comforters, volunteers taking part in medical research, and others imaged using medical radiological equipment whilst the latter relates to occupational and public exposure. A Medical Physics Expert often doubles up as the Radiation Protection Expert if he is qualified, indeed in small hospitals this is often the case.

3.8 Strong information and communication technology (ICT) skills

Today most medical devices are ICT based and the use of ICT pervades all areas of healthcare, yet the level of knowledge of the subject is in the case of many healthcare professionals limited to office productivity software (mainly word processing and presentation). Expertise in basic number crunching software such as spreadsheets and basic statistical packages, if any, is limited to the use of a few basic functions. Knowledge of ICT hardware is very limited and knowledge of the intricacies of hardware connectivity is extremely limited if not totally absent. Again, Medical Physicists have a very high skill level in ICT skills including advanced skills in programming and the use of high level packages such as 'MATLAB'.

3.9 High level qualification and curriculum frameworks and ethical standards

The qualification and curriculum frameworks leading to the certification of Medical Physics professionals in many countries are major tools (and in many countries enforced by legal requirements) in ensuring the high calibre of such professionals and the improved harmonisation of professional standards. To become a Medical Physicist one has to first go through a Bachelor's degree in physics (or equivalent), a Master's degree in physics or medical physics (or equivalent) and in Europe a minimum two-year supervised traineeship under the direction of a high level Medical Physicist in the specific area of specialization of the trainee. To reach the MPE level in Europe (which is at European Qualifications Level 8, the highest level of qualification level in Europe) requires a minimum further two years training and advanced experience in the particular speciality [5, 6]. This framework ensures that Medical Physics professionals have the necessary knowledge and skills to deliver effective and safe practice and that patients are entrusted to a safe pair of hands. Although the above framework was mostly developed in Europe (and the US [16, 17]) similar frameworks are recommended by the IAEA and the International Medical Physics Certification Board (IMPCB) for the international community [10, 18]. Regions and individual countries where such a framework does not exist aspire to have a similar one [19]. In addition, the profession prides itself on its high ethical standards [20, 21].

3.10 High level service standards and practices

The organizations representing Medical Physicists have set high standards for Medical Physics services. Examples of such standards are EFOMP Policy Statement 13 and British Standard BS 70000:2017 [22, 23]. In addition, the Medical Physics profession (and its sister profession biomedical engineering) sets the standards in the technical aspects of the use of medical devices in medicine. Many regional organizations like EFOMP and country organizations such as AAPM and the IPEM regularly publish technical reports which are in turn often used by international standard setting bodies such as the International Electrotechnical Commission to set international standards. Examples can be found on these sites[7].

References

[1] Caruana C J 2012 The ongoing crisis in medical device education for healthcare professionals: Breaking the vicious circle through online learning *Int. J. Reliable Qual. E-Healthcare* **1** 29–40

[2] Caruana C J *et al* 2018 EFOMP policy statement 16: The role and competences of medical physicists and medical physics experts under 2013/59/EURATOM *Phys. Med.—Eur. J. Med. Phys.* **48** 162–8

[3] Pitts P *OSH Wiki Physical Agents* https://oshwiki.eu/wiki/Physical_agents [accessed 20 December 2019]

[4] Berris T, Nüsslin F, Meghzifene A, Ansari A, Herrera-Reyes E, Dainiak N, Akashi M, Gilley D and Ohtsuru A 2017 Nuclear and radiological emergencies: Building capacity in medical physics to support response *Phys. Med.—Eur. J. Med. Phys.* **42** 93–8

[5] Caruana C J, Christofides S and Hartmann G H 2014 EFOMP Policy Statement 12.1: recommendations on medical physics education and training in Europe 2014 *Phys. Med.—Eur. J. Med. Phys.* **30** 598–603

[6] Caruana C J *et al* 2014 Qualification and curriculum frameworks for the medical physics expert in Europe - inventory of learning outcomes for the medical physics expert in Europe *European Guidelines on the Medical Physics Expert (Radiation Protection Series 174)* ed E Guibelalde, S Christofides, C J Caruana, S Evans and W van der Putten (Luxembourg: Publications Office of the European Union) ch 3, pp 15–20 and Annex I, pp 1–59.

[7] Caruana C J 2014 Learning outcomes in radiation protection for Medical Physicists and medical physics experts *MEDRAPET Guidelines on Radiation Protection Education and Training of Medical Professionals in the European Union (Radiation Protection Series 175)* (Luxembourg: Publications Office of the European Union: European Commission), ch 7, pp 73–5

[8] Sahiner B, Pezeshk A, Hadjiiski L M, Wang X, Drukker K, Cha K H, Summers R M and Giger M L 2019 Deep learning in medical imaging and radiation therapy *Phys. Med.—Eur. J. Med. Phys.* **46** 1–36

[9] IAEA 2014 *Radiation Protection and Safety of Radiation Sources: International basic safety standards: General Safety Requirements Part 3, No. GSR Part 3 Vienna*

[10] IAEA 2013 *Roles and Responsibilities, and Education and Training Requirements for Clinically Qualified Medical Physicists, IAEA Human Health Series No. 25*

[11] IAEA 2013 *Postgraduate Medical Physics Academic Programmes, Training Course Series No. 56*

[12] IAEA 2009 *IAEA Clinical Training of Medical Physicists Specializing in Radiation Oncology, Training Course Series No. 37*

[13] IAEA 2010 *IAEA Clinical Training of Medical Physicists Specializing in Nuclear Medicine, Training Course Series No. 50*

[7] AAPM https://www.aapm.org/pubs/reports/tabular.asp; IPEM(UK). https://www.ipem.ac.uk/ScientificJournals Publications/IPEMReportSeries/AvailablePublications.aspx; EFOMP. https://www.efomp.org/index.php?r=fc&id=protocols.

[14] IAEA 2010 *Clinical Training of Medical Physicists Specializing in Diagnostic Radiology, Training Course Series No. 47*

[15] Caruana C J *et al* 2014 The role of the medical physics expert *European Guidelines on the Medical Physics Expert (Radiation Protection Series 174)* ed E Guibelalde, S Christofides, C J Caruana, S Evans and W van der Putten (Luxembourg: Publications Office of the European Union: European Commission), ch 2, pp 11–4

[16] AAPM 2009 *Academic Program Recommendations for Graduate Degrees in Medical Physics, Report* No. 197

[17] AAPM 2013 *Essentials and Guidelines for Medical Physics Residency Training Programs, Report* No. 249

[18] *International Medical Physics Certification Board. A Model for the Medical Physics Certification Process* https://impcbdb.org/model-program/ [accessed 20 December 2019]

[19] ROUND W *et al* 2011 *AFOMP Policy Statement No. 3: Recommendations for the Education and Training of Medical Physicists in AFOMP Countries*

[20] Sherriff S B and Dendy P P 2003 Guidelines on professional conduct and procedures to be implemented in the event of alleged misconduct: EFOMP policy statement No 11 *Phys. Med.—Eur. J. Med. Phys.* **19** 227–9

[21] Skourou C *et al* 2019 mReport No. 109—code of ethics for the American Association of Physicists in Medicine (revised): report of task group 109 (2019) *Med. Phys.* **46** 79–93

[22] Christofides S 2008 The European Federation of Organizations for Medical Physics policy statement No. 13: Recommended guidelines on the development of safety and quality management systems for medical physics departments *Phys. Med.—Eur. J. Med. Phys.* **25** 161–5

[23] British Standards Institution 2017 *BS 70000:2017 Medical Physics, Clinical Engineering and Associated Scientific Services in Healthcare. Requirements for Quality, Safety and Competence.* https://shop.bsigroup.com/ProductDetail/?pid=000000000030323397

IOP Publishing

Leadership and Challenges in Medical Physics: A Strategic and Robust Approach
A EUTEMPE network book
Carmel J Caruana

Chapter 4

Internal WEAKNESSES of Medical Physics

Learning outcomes

By the end of the chapter the reader will be able to:

- discuss the main weaknesses of Medical Physics professionals.
- understand more deeply the meaning of the term 'critical weakness'.
- understand how crucial it is that weaknesses and particularly critical weaknesses be recognized and formally addressed.

4.1 Weaknesses and critical weaknesses

It is relatively easy to set up meetings to identify and discuss the strengths of a group. Everyone likes talking about what they are good at—however, weaknesses are another matter! We are all humans and we like to discuss issues that make us feel special but want to sweep under the carpet those that show up our deficiencies. However, not identifying, acknowledging and remedying our weaknesses is a fatal mistake. Only by admitting that a problem exists can we start the process of elimination of the problem. As Medical Physicists we are lucky that we are intelligent enough to be able to resolve most problems once we learn to accept and face them head on. CRITICAL WEAKNESSES are the opposite of core competences—whilst core competences ensure that a profession thrives, critical weaknesses are a disease that can destroy it from within unless treated. We present what we consider are the weaknesses of Medical Physics.

4.2 Critical weakness: absence of a universally acknowledged easily-marketable mission statement for the profession

One of the problems that the profession faces is the absence of a succinct, unambiguous universally acknowledged, and *very importantly easily-marketable*

doi:10.1088/978-0-7503-1395-7ch4

mission statement which would be a beacon for Medical Physics professionals across the world and that can be easily put into action to 'sell' the role to healthcare management and indeed society at large. Many Medical Physics professionals are at a loss when they are asked the question: 'But what do Medical Physicists do exactly?' We all know what we do but are unable to explain it succinctly. As strategy meetings with high level healthcare officials are of necessity very short affairs (as such officials are often so bogged down with many responsibilities) being able to communicate in a succinct manner (in a few minutes not hours!) the essence of our role is a critical skill. Unfortunately, Medical Physics leaders worldwide have given very scant attention to this issue. An absence of a clear mission statement leads to serious problems in marketing the profession and handicaps the development of vision statements as the latter need to be based on a strong mission statement for the profession. Decisions regarding role and scope of practice are similarly hampered as are discussions with other stakeholders. The absence of such a statement leads to ambiguity in the role and absence of harmonisation in the scope of practice.

Two fundamental questions need to be addressed before the mission statement can be finalised:

- should we limit our role to devices in diagnostic and interventional radiology, nuclear medicine and radiation oncology and ionizing radiation as it is from these areas that the legal basis for the profession emanates in many countries? This issue is further discussed in the following section as it is linked to another critical weakness which is *narrow range of specializations.*
- what is the distinction in the missions of Medical Physicist and biomedical engineers? Although these two professions are sister professions and there is a degree of overlap between the professions, their role is distinct with respect to orientation. Medical Physics is *patient oriented* and is mainly concerned with ensuring that existing devices are introduced safely into healthcare and used effectively, safely and efficiently. Their competences are moving ever closer to those of medical practitioners and healthcare professionals with increasing levels of knowledge of anatomy, physiology and pathology in the curriculum. Clinical biomedical engineers, on the other hand are more manufacturer, device-provider, installation and maintenance oriented. They often aim towards the further development of the technology. To be able to do this they need expertise in electronics, signal processing and medical device hardware proper. An initial discussion on the issue can be found here [1].

4.3 Critical weakness: narrow range of specializations

The range of specialties of clinical Medical Physics is in many countries still mostly restricted to the use of ionizing radiation in D&IR, NM and RO. Most countries also include occupational/public/environmental radiation protection (known as 'Health Physics' in the US) as a further specialty whilst others have made imaging with non-ionising radiation a separate specialty (although one can argue that it is really part of D&IR). It is our firm view that although these areas of specialization are very wide in scope, limiting the scope of the profession in this manner is risky as it is a case of putting all eggs in one basket. A profession that does not seek to expand its role will eventually contract owing to the pressures arising from commoditization. *It would be wise to extend our scope to all medical devices and all physical agents.* Some countries have already moved to expand the range of specialties e.g. audiology in the Netherlands, neurology and neurological physiology in Finland, physiological measurement in the UK. It is very important that Medical Physicists seek to expand their role in a proactive manner.

4.4 Critical weakness: insufficient strategic and robust leadership skills

The Medical Physics profession still has not realized the importance of strategic and robust leadership (and indeed this book is an attempt to resolve the issue). Few if any lectures on leadership are held during Medical Physics education and training programmes and leadership courses are practically non-existent. Fundamental to this weakness is an absence of subjects such as organizational psychology, organizational politics and negotiating skills. Indeed, these subjects are so crucial to leadership today that chapters 8–10 of this book are dedicated to them.

The AAPM has set up a 'Medical Physics Leadership Academy (MPLA) Committee' to start the ball rolling. Its charge at present (November 2019) can be found here[1].

Note that unfortunately terms such as 'strategic planning', 'strategic leadership', 'robust leadership' do not even feature in the charge description. There seems to be more an emphasis on business administration than strategic leadership. The Facebook page of the Academy is only followed by 31 people (accessed 9 November 2019). To be fair the Academy does offer a summer school, however, it is simply a three-day affair and it is difficult to envisage how one can convert non-leaders into leaders over a three-day period as leadership requires a long term transformation of the individual.

The only comprehensive module for leadership at present known to the author is that organized by the EUTEMPE network consortium in Europe (www.eutempe-net.eu/mpe01) and was the brainchild of the author of this book. The module involves 60 hours of online and 20 of onsite work, practical case studies in leadership and face-to-face discussions with European leaders of the profession. The author hopes that this book will stimulate leaders from other regions of the world to develop similar leadership modules of their own in their region.

4.5 Critical weakness: low marketing skills leading to too low profile of the profession within and outside healthcare

It is no use having high levels of expertise and a highly relevant mission statement if nobody knows about them! Many Medical Physicists seem to find it difficult or are unwilling to market, publicize and sell their services. This attitude contrasts sharply

[1] https://www.aapm.org/org/structure/?committee_code=MPLAWG.

with that of other healthcare professions who use all available opportunities to advertise their profession on an ongoing basis. In many countries educational programmes for Medical Physicists are organized within faculties of physics, science or engineering and or in technical universities so that healthcare professional students rarely if ever meet Medical Physics students. Hence most healthcare professions (with the exception of radiographers, radiologists, radiation oncologists and nuclear medicine physicians) have heard little of the profession (if any). The problem is even worse when it comes to the general public. The percentage of people who know about the profession is miniscule—which translates to low political power. Medical Physics organizations should hold marketing skills seminars on an ongoing basis and appoint outreach persons to raise public awareness of the profession. If the profession is to move forward, marketing of the profession should be carried out on a proactive and ongoing basis, particularly on the public media.

4.6 Critical weakness: insufficient regional/international networking

We live in an age of fast regionalisation and indeed globalisation. Yet the number of Medical Physicists who are ready to dedicate some of their time to regional/international networking particularly with regard to political/professional/educational issues is still much too low. This has led to several effects ranging from non-harmonised conception of the role itself to non-harmonised standards of practice and insufficient access to funding. Medical Physicists need to understand more fully that in today's globalised world, regional and international networking is not simply a nice-to-have-but-not-essential matter but a crucial means of increasing the political strength of the profession.

4.7 Critical weakness: low level of preparation for the realities of healthcare organizational politics

The organizational politics of healthcare establishments are very different from those of a faculty of science. A faculty of science is organized along well-defined departments which are quite mutually exclusive: physics, chemistry, biology. Every student knows his place in the system so that a physics graduate brought up in such an environment leads a relatively calm and sheltered existence away from inter-professional conflict. Transfer such a graduate to the hospital environment and they are faced with a completely different set of rules in terms of sociology, organizational psychology and politics. Instead of a few well-defined departments and areas of study they are faced with a gigantic organization consisting of a multitude of inter-connected departments. Instead of a well-defined role with next to no competition from other sciences they find themselves immersed in a multi-professional organization consisting of tens of different healthcare professions jockeying for power (under the euphemism of 'role-development' which is often synonymous with role-poaching). This is not helped by the role overlaps which more often than not lead to inter-professional strife. Again students of faculties of science are usually quite homogeneous in the sense that they are similar in level of intelligence and motivation. In a hospital one has to deal with staff of all levels from low staff to medical specialists: different psychologies, different attitudes, and different chips-on-shoulders. The situation sometimes becomes quite stressful for young Medical Physics trainees. We as educators and trainers have an obligation to prepare our trainees not only scientifically but also psychologically for such an environment.

4.8 Insufficient number of independent departments

The location of the Medical Physics group within the organizational structure of a healthcare organization has an impact on the possibilities available for role-development and strategic planning. Unfortunately, there are still too many hospitals without an independent Medical Physics department and in such cases Medical Physicists are often distributed in various clinical departments such as D&IR, NM and RO and with little contact between the Medical Physicists in the various departments. When Medical Physics professionals are organized within a single department there is a strong element of mutual-support and cross-fertilization of ideas across the different specialties. In those hospitals in which Medical Physicists are dispersed in different clinical departments the role may become fragmented leading to isolated professionals with little political influence on departmental affairs and therefore little impact on clinical services.

4.9 Non-harmonised scope of practice for the profession

Owing to the various explicit or implicit mission statements the scope of practice of the Medical Physics profession globally, regionally and in the case of large states even nationally is not harmonised. This is having an impact not only on the profession but also on patient safety. Hospitals with well-established Medical Physics services lead to high quality care whilst those which lack such services lead to lower healthcare service quality and higher rates of morbidity[2] and mortality.

4.10 A reluctance by some Medical Physicists to be part of the wider healthcare picture

Owing to the emphasis on the contribution of physics to the development of modern society during undergraduate physics programmes, there is a tendency by some Medical Physicists to consider that physics is what is really important, that other subjects (particularly the soft subjects) are less important and there is little need for Medical Physicists to bother about them at all. This assertion is a complete fallacy! Whether a patient's health is improved or otherwise is a not only a function of physics input but also the result of health economics, level of inter-professional understanding and teamwork, level of knowledge and skills of the roles of the various healthcare professions, ethical issues, education and training, management and communication among others. If Medical Physicists wish to achieve a higher

[2] Definition: 'Refers to having a disease or a symptom of disease, or to the amount of disease within a population. Morbidity also refers to medical problems caused by a treatment' (https://www.cancer.gov/publications/dictionaries/cancer-terms?expand=M).

level of involvement in healthcare they need to look at the wider picture and consider their role within that wider picture. Such subjects need to be included in our education and training and we need to learn the terminology and jargon used across healthcare, we need to build bridges to others. Self-imposed isolation is an unequivocal way of shooting oneself in the foot and being seen as impractical and overly theoretical or perhaps even arrogant.

4.10.1 Self-reflection/discussion point

If you are going to be a Medical Physics leader you would need to interact with other healthcare professions and hospital managers. You will need to understand and when necessary use the same terminology as they use. Let us test your appreciation of the wider context in which Medical Physicists work. In our work we use the term 'quality' quite often. Consider the phrase 'dimensions of quality healthcare'. Make a list of what you consider are the dimensions of quality healthcare and define each of them. Compare your list to that given in the following reference [2]. How many of them did you guess correctly? Which did you leave out? Can you link each of them with aspects of Medical Physics practice?

4.10.2 Self-reflection/discussion point

Medical and other healthcare professionals use the terms 'ethical' and 'not ethical' quite often. Can you list the four fundamental principles of healthcare ethics? Compare your list to the actual list in the following footnote[3]. Can you apply these to Medical Physics work? You can read about their application in radiation medicine here [3].

[3] BENEFICENCE–NONMALEFICENCE: beneficence means that our actions are aimed to do good to others, in this case the patient. Healthcare professionals have a primary obligation to use the best available healthcare technology to promote the well-being of patients by preventing or curing diseases, relieving suffering and improving health status. Healthcare professionals must be benevolent in their actions. Nonmaleficence means that while maximizing the patient's well-being, healthcare professionals should minimize risks and deleterious effects to the patient.
PRUDENCE–REASONABLENESS: when faced with uncertainties in risk estimates apply the precautionary principle but avoid excessive unreasonable counter-productive zeal in reducing risk.
RESPECT FOR HUMAN DIGNITY: there are many aspects to patient dignity. Certainly as healthcare professionals we must respect patients' autonomy. We should completely and honestly inform patients, safeguard confidentiality within the boundaries of law and empower patients to make informed decisions about diagnostic and therapeutic procedures. Patients' decisions about their care must be paramount, as long as those decisions are in line with ethical practice and do not lead to demands for inappropriate care. Any diagnostic or therapeutic procedure demands the obligation to provide information to the patient and the obtaining of informed consent. The value of respect for the dignity of the individual means that the patient must be involved in decisions affecting him.
SOCIAL JUSTICE: healthcare professionals should promote justice in the healthcare system including the fair distribution and cost-effective use of limited healthcare resources and elimination of discrimination in healthcare, whether based on age, gender, sexual orientation, race, religion, socioeconomic status or any other social category.

4.10.3 self-reflection/discussion point

Healthcare management and professionals use the term 'healthcare economics' quite often. Can you define what we mean by 'economics'? Compare your definition to the one in the following footnote[4]. Reflect on the impact of health economics on Medical Physics work.

4.11 Insufficient communication and pedagogical skills

Our role as Medical Physicists includes the responsibility of teaching not only Medical Physics trainees but also members of the other healthcare professions at various levels from undergraduate to specialist level. In the case of radiation protection and in Europe this is a requirement of EU law[5]. Yet many Medical Physicists do not give sufficient importance to honing their communication skills. Many still are of the opinion that knowing a topic is sufficient to be able to communicate it to others and they give scant attention to preparing a good presentation. Many still do not realize that preparing a presentation for radiographers or radiologists simply by watering down one originally meant for Medical Physicists is simply not the way to go. The backgrounds and psychologies are simply different. In particular, the same topic but for different professions needs to be reformulated *from the perspective of the learning needs and specific role of that particular profession.*

4.12 Low level of research on professional and educational issues—low qualitative research methodology skills

All healthcare professions today have realized the importance of research on educational and professional issues, yet in the case of Medical Physics, research in these areas has barely taken off. In some Medical Physics journals such articles are even discouraged as they are not considered to be 'scientific enough' and therefore practically worthless or even the result of an inability of the researchers to do 'real' research. This is of course narrow thinking at its best and what has led to the problems we now face as a profession. Strategic planning itself needs to be based on real world research. One cannot produce a good strategic plan without real-world data on say the state of the profession at a particular point in time. Allied to this is the fact that such research often requires a strong element of *qualitative* research techniques (e.g. interviews, focus groups, Delphi technique) or *mixed quantitative–qualitative methods* and few Medical Physicists are capable of using such methods. The situation is slowly changing but needs to be accelerated [5, 6].

---

[4] Economics is concerned about how society allocates limited resources (in this case healthcare resources) among alternative uses/users: Which healthcare services shall we provide? How shall we produce them? Who shall receive them?

[5] EU Dir 2013/59/EURATOM Article 83(2)(h) [4].

References

[1] van der Putten W, Smith C E and Orton C G 2012 Point/Counterpoint. The professions of Medical Physics and Clinical Engineering should be combined into a single profession 'Clinical Science and Technology' *Med. Phys.* **39** 2953–5

[2] WHO 2006 *Quality of Care—A Process for Making Strategic Choices in Health Systems* https://apps.who.int/iris/bitstream/handle/10665/43470/9241563249_eng.pdf?sequence=1&isAllowed=y [accessed 20 December 2019]

[3] Cho K W, Cantone M C, Kurihara-Saio C, Le Guen B, Martinez N, Oughton D, Schneider T, Toohey R and ZöLzer F 2018 ICRP Publication 138: ethical foundations of the system of radiological protection *Ann. ICRP* **47** 1–65

[4] European Council 2013 *Council Directive of 5 December 2013 Laying Down Basic Safety Standards for Protection against the Dangers Arising from Exposure to Ionising Radiation, and Repealing Directives 89/618/Euratom, 90/641/Euratom, 96/29/Euratom, 97/43/Euratom and 2003/122/Euratom, Official Journal of the European Union L-13 of 17/01/2014.*

[5] Caruana C J, Wasilewska-Radwanska M, Aurengo A, Dendy P P, Karenauskaite V and Malisa M R *et al* 2010 A comprehensive SWOT audit of the role of the biomedical physicist in the education of healthcare professionals in Europe *Phys. Med.—Eur. J. Med. Phys.* **26** 98–110

[6] Castillo J, Caruana C J, Morgan P S, Westbrook C and Mizzi A 2019 Development of an inventory of biomedical imaging physics learning outcomes for MRI radiographers *RadioGraphy* **25** 202–6

Chapter 5

External environmental OPPORTUNITIES for Medical Physics

Learning outcomes

By the end of the chapter the reader will be able to:

- discuss the main opportunities available for the further development of the Medical Physics profession;
- understand more deeply the meaning of the term 'key opportunity';
- understand that strategic planning for Medical Physics should include a focus on the grasping of opportunities to further our strengths and reduce or completely eliminate weaknesses and threats;
- understand the importance of the fact that every opportunity lost by us is an opportunity for our competitors.

5.1 The importance of grasping opportunities

In this chapter we discuss the main opportunities in Medical Physics. KEY OPPORTUNITIES are those opportunities with higher benefit value. Strategic planning involves also the grasping of existing or expected opportunities (and whenever possible the creation of new opportunities) for the further strengthening of strengths and the minimization or complete elimination of weakness and threats with particular attention being given to core competences, critical weakness and existential threats. We also suggest some ways in which these opportunities can be used to advantage. *One cannot but overemphasize the importance of grasping opportunities and to realize that every opportunity lost is an opportunity for our competitors.*

doi:10.1088/978-0-7503-1395-7ch5

5.2 Key opportunity: ever increasing number and sophistication of hospital medical devices

The ongoing rapid expansion in the number and sophistication of medical devices is an opportunity that needs to be exploited fully by Medical Physicists. The level of expertise of the Medical Physics profession (and when clinically oriented its sister profession Biomedical Engineering) in this area exceeds by far the level of expertise of any other medical or healthcare profession. Therefore, the ever increasing number and sophistication of medical devices is an opportunity that matches perfectly our core competences. In particular, in our traditional areas of excellence—medical imaging and radiation oncology—new and more sophisticated devices and upgrades are the order of the day. Artificial intelligence, machine learning and theranostics are making huge headway, yet very few healthcare professions have an idea what the terms mean, let alone are capable of acceptance testing and quality controlling them or optimising their use. The expansion and increase in sophistication of medical devices is not limited to these areas—physiological measurement, use of lasers and biomedical optics, nanotechnology are all on the rapid increase and the Medical Physics profession should consider seriously expanding its role in these areas. This is an opportunity that is perfectly aligned to our strengths—and at the same time it is the weakness of our competitor healthcare professions. The healthcare industry is faced with healthcare professionals with insufficient physical and mathematical knowledge and skills to understand and optimize the devices they use on a daily basis. In addition, things are becoming more problematic as technology roars further ahead whilst the education and training of healthcare professions lags behind owing to the inertia of the educational system and reduced educational budgets [1]. The results of this contradiction are there for all to see: expensive equipment which is used at a fraction of its capability, less than effective use of medical devices and increasing number of safety issues, accident and incidents [2–4]. In addition, a substantial percentage of clinical protocols have in the past been based more on anecdotal evidence, 'expert' opinion and sometimes try-it-and-hopefully-we are-doing-the-right-thing approach and less on scientific optimization. Meanwhile the evidence-based-medicine movement (also known as evidence-based-healthcare) is an attempt to rectify this situation and put clinical practice on a more scientific basis. This requires professionals with a very strong understanding of the workings of medical devices and with a high level of research expertise. *Medical Physicists are best placed to do this and should use every available opportunity to promote themselves as experts in the scientific, clinically-effective, safe and efficient use of medical devices.*

5.3 Key opportunity: increased public awareness of the need for the establishment and maintenance of quality and patient safety standards in healthcare

The condition of too many patients deteriorates in hospitals owing to the low level of expertise regarding medical devices within these institutions. Social and political pressure for improved standards is continuously on the increase as the internet makes hospital incidents and accidents[1] mainstream news [4]. With regards to safety the publication of several reports highlighting incidents of adverse events within clinical settings has led to a heightened awareness that patient safety standards in hospitals are far from satisfactory. A report of the Institute of Medicine (US) has put the number of deaths through medical error in US hospitals at up to 98 000 per year [5]. Even in countries with well-developed healthcare systems such as the United Kingdom, it has been estimated that several hundred patients are killed or seriously injured per year from medical device incidents [6]. The stress experienced by healthcare professionals arising from fear of being unable to use medical devices properly is well documented [7, 8]. In response, the EU in its 'Luxembourg Declaration on Patient Safety' of the 5th of April 2005 made it clear that: 'Access to high quality healthcare is a key human right recognized and valued by the European Union, its institutions and the citizens of Europe. Accordingly, patients have a right to expect that every effort is made to ensure their safety as users of all health services' [9]. The EU further recommends that a safety culture needs to be established within hospitals. A major opportunity for the Medical Physicist as an educator of healthcare professionals is embedded in the following recommendation from the same document: 'To recognize and support the user training provided by medical devices, tools and appliances manufacturers thereby ensuring the safe use of new medical technology and surgical techniques'. *It is recommended that Medical Physics organizations link up with patient organizations and project themselves as patient advocates regarding safety. Their websites should be sources of information about medical devices and physical agents. They should inform the general public of incidents and accidents.*

5.4 Key opportunity: heightened public awareness regarding occupational safety and environmental issues

Increased awareness of occupational safety and environmental issues is another social phenomenon which is on the increase. Healthcare workers are expected to take responsibility for their own personal safety and the safety of colleagues whilst the public is becoming more aware of the deleterious effect of some physical agents in the environment. The inclusion of competences concerning protection from physical agents within the education and training curricula of healthcare professionals is a legal requirement and constitutes an excellent opportunity for the Medical Physicist with an interest in education and training. *It is recommended that Medical Physics organizations link up with occupational health and safety organizations and environmental groups and project themselves as occupational/public advocates with regard to safety from physical agents. The general public too needs to be educated and our websites should be sources of information about such issues. We should inform worker organizations/general public of incidents and accidents and publicise how the presence of Medical Physicists would have led to the avoidance of such incidents and accidents.*

[1] An incident is a 'near miss' where nobody got hurt; in an accident somebody got hurt.

5.5 Key opportunity: ever expanding and developing legislation

In many countries in the world, legislation with respect to the clinically-effective, safe and efficient use of medical devices for patients is on the increase, as is legislation concerning the protection of workers and the public from the detrimental effects of physical agents. The EU has promulgated several directives regarding medical devices, safety from physical agents and personal protective equipment which are binding on all states. Two directives[2] cover medical devices (including radiation aspects) [10, 11]. Major opportunities are also offered by directives regarding protection from physical agents, a list of which can be found here [12]. Increasingly, regulations make mandatory radiation protection education for those healthcare professionals who may be exposed to ionizing radiation, who expose patients for medical reasons or who may act as prescribers for medical images. Other directives concern occupational exposure to vibration (relevant to dentistry), electromagnetic radiation (very relevant to MRI) and artificial optical radiation (very relevant to say uses of lasers in medicine) [12]. Nowhere is the expansion of legislation so manifest as in the case of ionising radiation. The IAEA Basic Safety Standards in the use of ionising radiation [13] and their equivalent in the various regions and countries of the world such as EU directive 2013/59/EURATOM [14] are models of how legislation should be developed to ensure safety for all. This legislation has been of immense importance to the Medical Physics profession as it has helped put the profession on a sound legal basis. Indeed, both the IAEA and in the case of Europe the European Commission have provided substantial funds for the development of Medical Physics education and training. In the EU the European Commission has helped EFOMP develop the role and curricula for Medical Physicists and made funds available for the advanced education and training of Medical Physicists[3]. *Such funds have provided an immense boost to the profession and are a key opportunity that should be appreciated and utilised to the full.* In addition, such legislation offers opportunities for role expansion [15].

However, this is not enough. *The Medical Physics profession should create further opportunities for itself by exerting pressure on legislators to recognize that patient safety through the effective use of medical devices and protection from physical agents should not be limited to ionising radiation and strive to ensure that the rigorous standards of safety to be found in ionising radiation are extended to other areas such as magnetic resonance imaging, lasers and in the future nanodevices. For example, owing to pressures from unions the EU has promulgated legislation for occupational safety regarding electromagnetic radiation for workers but none for patients. This is an*

[2] The term 'directive' is increasingly being replaced by the term 'regulation'.

[3] For example look at the project European Training and Education for Medical Physics Experts (www.eutempe-net.eu) which developed a set of education and training modules to turn newly certified Medical Physicists into Medical Physicists at expert level (known as Medical Physics Experts in EU legislation) in diagnostic and interventional radiology. Again the ENEN+ project (https://plus.enen.eu/description/) provided money for young Medical Physicists to attend the courses and conferences necessary to improve their careers. Both were funded by the European Commission.

opportunity for Medical Physicists to help create awareness among the general public about such issues hence projecting the profession into the public sphere and making it better known.

5.6 Key opportunity: escalating cost of healthcare, the need for efficient use of medical devices and Heath Technology Assessment

Health economics is the science of making optimum use of scarce resources and is particularly relevant as the cost of healthcare continues to escalate and the expectations of patients become higher. For the purpose of this work health economics is most relevant with respect to the economical use of devices and the need for health technology[4] assessment. The WHO defines Health Technology Assessment (HTA) as the 'systematic evaluation of properties, effects and/or impacts of health technologies and interventions. It covers both the direct, intended consequences of technologies and interventions and their indirect, unintended consequences' and 'is used to inform policy and decision-making in health care, especially on how best to allocate limited funds to health interventions and technologies'[5]. *The relative effectiveness of the different technologies we use is increasingly subject to debate* [16] *and we as Medical Physicists need to take this opportunity and involve ourselves in such issues, as it would raise our profile with health economists.* An initial discussion on the role of the Medical Physicist in clinical trials (often included in HTAs) can be found here [17].

5.7 Key opportunity: the rapid expansion in the number of home-use, self-testing and wearable devices

A home-use medical device is a medical device that is 'intended for users in any environment outside of a professional healthcare facility'[6]. A self-testing device is any 'device intended by the manufacturer to be used by lay persons' [11]. Wearable technology refers to devices used to collect data for monitoring a user's health such as: heart rate, calories burned, release of certain biochemicals, epilepsy, stress, electrocardiogram and risks of age-dependent disease. The use of such devices has mushroomed over the last few years as more and more patients prefer to stay in private home care facilities or indeed their own home as opposed to large hospitals. This tendency will increase as problems of inadequate healthcare provision and infection control in hospitals increasingly hit the media. Such devices are not only a clinical opportunity for Medical Physicists *but also can be turned into a business opportunity. One can have a vision of shops selling such devices manned by Medical*

---

[4] 'A health technology is the application of organized knowledge and skills in the form of devices, medicines, vaccines, procedures and systems developed to solve a health problem and improve quality of lives' (WHO, https://www.who.int/health-technology-assessment/about/healthtechnology/en/). Note that medical devices are only one category of healthcare technology.

[5] https://www.who.int/health-technology-assessment/about/en/.

[6] https://www.fda.gov/medical-devices/home-health-and-consumer-devices/home-use-devices.

Physicists who can give advice to customers on the most appropriate devices to purchase, proper and safe use, re-calibration frequencies and other information. Medical Physicists can also offer quality control services for such devices.

5.8 Key opportunity: need for technically oriented people on modern hospital governance boards

As medical device technology and protection from physical agents take an ever increasing role in healthcare there is an increasing need for more persons with relevant expertise on hospital governance boards. Medical Physicists are the ideal candidates to provide expert advice in these areas. *Medical Physicists should endeavour to become members of such boards as they offer immense opportunities to make the profession better known in the hospital.* EFOMP has put out a policy statement in this regard [18]. In this policy statement EFOMP encourages Medical Physicists to:

(a) Seek appointment on their hospital's governance board;
(b) Participate actively in their hospital's governance board committees;
(c) Promote the Medical Physics profession to their hospital's governance board and its committees.

EFOMP also recommends that all national Medical Physics professional organizations should 'encourage their Medical Physicists to be closely involved in hospital governance and, where this has not already happened, to seek membership of their hospital's governance boards and its committees, emphasising the importance of such membership for the good of the patient and the hospital as a whole'.

References

[1] Caruana C J 2012 The ongoing crisis in medical device education for healthcare professionals: Breaking the vicious circle through online learning *Int. J. Reliable Qual. E-Healthcare* **1** 29–40
[2] Tamarat R and Benderitter M 2019 The medical follow-up of the radiological accident: Épinal 2006 *Radiat. Res.* **192** 251–7
[3] Peiffert D, Simon J M and Eschwege F 2007 Epinal radiotherapy accident: past, present, future *Cancer Radiother.* **11** 309–12
[4] Chustecka Z 2013 (February 6th) Docs in prison after radiation overdose in prostate cancer *Medscape Medical News*
[5] Kohn L T, Corrigan J M and Donaldson M S (ed) 2000 *To Err is Human: Building a Safer Health System* (Washington, DC: Committee on Quality of Health Care in America, Institute of Medicine)
[6] Amoore J and Ingram P 2002 Learning from adverse incidents involving medical devices *Br. Med. J.* **325** 272–5
[7] Almerud S, Alapack R J, Fridlund B and Ekebergh M 2008 Beleaguered by technology:care in technologically intense environments *Nurs. Philos.* **9** 55–61
[8] Barnard A 2000 Alteration to will as an experience of technology and nursing *J. Adv. Nurs.* **31** 1136–44

[9] *European Commission Patient Safety—Making it Happen! Luxembourg Declaration on Patient Safety of the 5th of April 2005*
[10] *REGULATION (EU) 2017/745 OF THE EUROPEAN PARLIAMENT AND OF THE COUNCIL of 5 April 2017 on medical devices, amending Directive 2001/83/EC, Regulation (EC) No 178/2002 and Regulation (EC) No 1223/2009 and repealing Council Directives 90/385/EEC and 93/42/EEC*
[11] *REGULATION (EU) 2017/746 OF THE EUROPEAN PARLIAMENT AND OF THE COUNCIL of 5 April 2017 on in vitro diagnostic medical devices and repealing Directive 98/79/EC and Commission Decision 2010/227/EU*
[12] Pitts P *OSH Wiki Physical Agents* https://oshwiki.eu/wiki/Physical_agents [accessed 20 December 2019]
[13] IAEA 2014 *Radiation Protection and Safety of Radiation Sources: International Basic Safety Standards: General Safety Requirements Part 3, No. GSR Part 3 Vienna*
[14] European Council 2013 *Council Directive of 5 December 2013 Laying Down Basic Safety Standards for Protection Against the Dangers Arising from Exposure to Ionising Radiation, and Repealing Directives 89/618/Euratom, 90/641/Euratom, 96/29/Euratom, 97/43/Euratom and 2003/122/Euratom, Official Journal of the European Union L-13 of 17/01/2014*
[15] Caruana C J, Tsapaki V, Damilakis J, Brambilla M, Martín G M, Dimov A, Bosmans H, Egan G, Bacher K and McClean B 2018 EFOMP policy statement 16: The role and competences of medical physicists and medical physics experts under 2013/59/EURATOM *Phys. Med.—Eur. J. Med. Phys.* **48** 162–8
[16] Chen A B 2014 Comparative effectiveness research in radiation oncology: assessing technology *Semin. Radiat. Oncol.* **24** 25–34
[17] Kron T 2013 The role of medical physicists in clinical trials: More than quality assurance *J. Med. Phys.* **38** 111–4
[18] Christofides S and Sharp P 2015 The European Federation of Organisations for Medical Physics Policy Statement no. 15: Recommended guidelines on the role of the medical physicist within the hospital governance board *Phys. Med.—Eur. J. Med. Phys.* **31** 201–3

Chapter 6

External environmental THREATS for Medical Physics

Learning outcomes

By the end of the chapter the reader will be able to:

- discuss the main threats to the development of the Medical Physics profession;
- understand more deeply the meaning of the term 'existential threat';
- understand the importance of acknowledging the existence of these threats in a formal manner;
- discuss the importance of researching threats and countering them in a proactive manner.

6.1 Threats are for elimination

All professions in this world face threats to their further development. This is part of the nature of things and not an issue specific to Medical Physics. The threats may arise from developments in the external environment or unfair competition between professions. It is important to understand that some threats are more critical than others and some threats are so serious that they pose a threat to the very existence of a profession (EXISTENTIAL THREATS). This is a fact of life that all mature, strategic and robust leaders recognize and accept as a challenge. *Such threats should be countered in the same manner that physicists counter all other problems: they should be acknowledged, researched, analysed, addressed and eliminated—and where they cannot be eliminated totally their impact must be proactively reduced to a safe level.*

doi:10.1088/978-0-7503-1395-7ch6

6.2 Existential threat: low number of physics and engineering graduates

Traditionally in many countries students come to Medical Physics with an undergraduate degree in physics or engineering (or equivalent). For various reasons, in many countries of the world the number of graduates in these areas is decreasing (and in some cases plummeting). However, many Medical Physics masters programmes depend on having a steady stream of physics/engineering graduates. The problem of course is that academics from such programmes are not encouraging their undergraduates to opt for a master's in Medical Physics as they barely have enough students for their own masters programmes. The result of all this is that there has been in recent years a dip in Medical Physics master's students and this means fewer Medical Physicists. Many countries are looking for ways of solving this issue. Some countries have started a combined Bachelors–Master's degree in Medical Physics whilst others have a combined Physics–Medical Physics or even Physics–Medical Physics–Radiation Protection undergraduate degree. *What is important is that when physics and engineering students are in short supply we must recruit our own undergraduate students and should not depend on others for students. We must ensure that the capability of recruiting talented students to the profession is totally within our control and is independent of the issues facing physics or engineering per se. Medical Physics as an area of study and as a profession is fascinating enough to attract students independently of physics or engineering.*

6.3 Existential threat: austerity economics and unrestrained commoditization

These twin inter-related threats have already been introduced in chapter one, however, because of their red-alert nature we reiterate and debate them further.

Austerity economics aims to reduce government budget deficits through spending cuts, tax increases, or a combination of both. The first targets of spending cuts are often public education, social services and of course public healthcare as they are areas of social activity without a direct incoming revenue stream and affect the rich (and often politically influential) members of society the least as the latter can always pay for private services. When healthcare budgets are reduced, money is diverted to issues with higher political impact. Quality in healthcare and education is often the first to suffer when austerity comes into play.

Commoditization refers to the process by which parts of services offered by high level professionals such as Medical Physicists and radiologists become cheaper to provide by establishing standard written protocols which permit such services to be carried out by lower level (and hence lower-paid) professionals (e.g. basic chest reporting done by specially trained radiographers instead of radiologists or in the case of Medical Physics establishing written daily QC protocols which can be carried out by Medical Physics assistants or radiographers) [1]. In modern societies, the ability to commoditize anything is seen as a benefit to all, and opens up higher level human resources that can be put to better use for service development and innovation (e.g. Medical Physicists having more time to focus on higher order QC tasks as opposed to spending time on routine simple quality control tasks). However, if not managed well, commoditization would ultimately pose a threat to the profession and quality of patient service e.g. radiology department managers under pressure to reduce costs opting to save money by employing a lower paid worker to do daily quality control who is then also tasked to do advanced quality control even though he/she is not capable—with the result that patient safety degrades; or radiographers being employed to read chest radiographs without taking the necessary precautions to ensure that difficult cases are automatically transferred to chest radiologists leading to increases in the number of misdiagnoses. Such commoditization has other effects and often problems of equity arise. Rich patients would be able to afford radiologists, reducing the risk of misdiagnosis whilst other patients are not. Ethical questions arise: in the case of public institutions, who would decide whether a patient is directed towards a fully qualified experienced radiologist as opposed to a radiographer and on what criteria? Are there issues of social justice involved? Patients attending large radiation oncology centres with well-developed and fully staffed Medical Physics departments get accurate and precise treatment; patients where there is an insufficient number of Medical Physicists have a higher rate of mortality and are victims of serious incidents [2–4]. This is particularly common in those departments where managers are uninformed of the services offered by radiologists and Medical Physicists. *Unbridled commoditization if not tackled well will destroy the high level professions and, once destroyed, who can bring them back*?

A related issue we must tackle is that although the availability of Medical Physics services is essential for quality and safe healthcare and often a legal requirement, such services do not have their own direct revenue stream. In situations where revenue becomes dominant, quality and safety are often the first victims (who care as long as patients do not notice it and do not take us to court? Who cares as long as it does not end up in the newspaper headlines?). *It is the opinion of the author that the only way to reduce the effects of these threats is for professions such as Medical Physics to take the lead in raising public awareness of the issues involved. The issue needs to be taken to the public to create political pressure and ensure that austerity leaning politicians keep their hands off healthcare quality and safety.*

6.3.1 Reflection/discussion point

Consider the following article which is an example of a politically driven commoditization-based research article:

Woznitza N, Piper K, Burke S and Bothamley G 2018 Chest x-ray interpretation by radiographers is not inferior to radiologists: a multireader, multicase comparison using JAFROC analysis *Acad. Radiol.* **25** 1556–63 [5].

The article is 'open access' which means that the authors have paid the publisher for it to be freely available for download to ensure maximum political impact. A cursory reading of the title would lead to the conclusion that one can replace *any* radiologist by *any* radiographer for chest diagnosis with no effect on patient safety.

The message it is sending to the financial management of healthcare institutions is basically: 'if you want to reduce costs all you have to do is simply replace all chest radiologists with any radiographers'. Before reading the article, consider the research methodology that should be used to provide concrete research evidence for the assertion in the title. Now read the article and consider whether the blanket statement made about radiographers' capabilities with respect to radiologists is supported by the methodology. Managers rarely have the time to read articles in full, many just read titles—do you realize the possible dangers such articles may pose to patient care and inter-professional relations? Consider the ethics of it all.

6.4 Existential threat: role poaching from other professions

Some members of other professions consider that they have a 'right' to poach the role of other professions under the umbrella term 'role expansion'. This is happening also in the case of Medical Physics where some professions consider it fair game to steal part of our role. Since we are better than they are and than they will ever be in the physics based aspects of healthcare and since they know they could never win in a fair competition they invariably tend to use unfair subterfuge to achieve their aims. Thankfully, few members of these professions are like this, however, unfortunately those few are quite vocal. It is a psychology driven by low creativity and emotional intelligence levels, egoism, envy and chips-on-shoulders. These people can produce a lot of damage both to the profession and to inter-professional relationships and hence the clinical service and patients if not recognized and stood up to.

References

[1] Fontenla D P and Ezzell G A 2016 Medical physicist assistants are a bad idea *Med. Phys.* **43** 1–3

[2] Tamarat R and Benderitter M 2019 The medical follow-up of the radiological accident: Épinal 2006 *Radiat. Res.* **192** 251–7

[3] Peiffert D, Simon J M and Eschwege F 2007 Epinal radiotherapy accident: past, present, future *Cancer Radiother.* **11** 309–12

[4] Chustecka Z 2013 (February 6th) Docs in prison after radiation overdose in prostate cancer *Medscape Medical News*

[5] Woznitza N, Piper K, Burke S and Bothamley G 2018 Chest x-ray interpretation by radiographers is not inferior to radiologists: a multireader, multicase comparison using JAFROC analysis *Acad. Radiol.* **25** 1556–63

Chapter 7

Healthy leadership and leadership styles

Learning outcomes

By the end of the chapter the reader will be able to:

- discuss the meaning of the terms healthy leadership, leadership style and corporate culture;
- understand that to be effective and motivational leadership needs to be people-oriented;
- discuss the characteristics of people-oriented leadership;
- discuss the various leadership styles;
- apply the use of various leadership styles in the different stages of development of a project team.

7.1 To be effective and motivational leadership needs to be people-oriented

Effective leaders know that they cannot achieve objectives on their own or without their team no matter how high their own personal level of intelligence or commitment. In today's complex multi-disciplinary healthcare milieu, in order for a Medical Physics team to be successful the different members of the team often have to take on specialised aspects of Medical Physics work, e.g. programming, 3D printing, use of advanced software such as MATLAB, ImageJ. The leader also knows that objectives can only be attained if each member of the team commits to them. The leader cannot oversee his team all the time, people have to work without supervision, do that little extra which is outside what they are strictly paid for, sometimes work in the evening and on the weekend. It is ultimately the team members who produce results: if the members of the team are motivated success is guaranteed, if not failure is the result. This is why an effective leader invests a substantial fraction of his/her time giving attention to the personal and professional

doi:10.1088/978-0-7503-1395-7ch7

needs of their team members—leadership needs to be people-oriented because without quality team members objectives will simply not be achieved.

7.2 What do we mean by 'healthy leadership'?

Effective teams lead to happy teams right? No!—research shows that it is the reverse—happy teams lead to effective teams! Happiness drives success—rather than the reverse [1]. Healthy leaders make sure that team objectives are not achieved at the expense of the health and happiness of the individual members in their team, nor indeed even at the expense of their own happiness and health. A healthy happy environment leads to increased motivation and creativity—unhealthy environments diminish productivity and destroy creativity. Unhealthy leaders destroy members of the team and ultimately destroy themselves. Fortune magazine's '100 best companies to work for' annually lists companies in order of employee happiness and satisfaction. In the years 2006–17 Google made it to the top of the list eight times. Its motto is: 'To create the *happiest*, most productive workplace in the world' and the vice-President of their department for people development has been quoted as stating that: 'It's less about the aspiration to be No. 1 in the world, and more that we want our employees and future employees to love it here, because that's what's going to make us successful' [2]. A people-oriented approach to leadership is crucial for team motivation.

7.3 Characteristics of people-oriented leadership

People-oriented leadership focusses on relationships and not just results. Healthy leadership focusses on creating a psychological environment conducive to good working and personal relationships between team members. There is a devotion of quality time to help build relationships based on an appreciation of and respect for others. The following are the core characteristics of people-oriented leadership.

7.3.1 Commitment to personal and staff development

A good leader knows that as circumstances and the environment change, both his staff and himself would need to adapt to those changes in order to cope and hence achieve their objectives. To adapt means one has to embrace new competences which means learning new personal and professional knowledge and skills. Most of this new learning in the case of Medical Physicists is often self-learning; however, it is also important to recognize that formal courses from intra-mural or extra-mural experts are also necessary and the necessary resources should be made available.

7.3.2 Respect for teamwork

In healthcare, as in so many other organizations, teamwork is a fundamental part of delivering successful outcomes. Teams work best when they understand their objectives, when roles are clear, and when individual members of the team are both respected for their contribution and trusted to do what is required. Objectives are developed with and by the team, and success is defined at the team level. When the psychological environment and behaviour within the team matches these characteristics good things happen. Research on teams shows that what really matters was less about who is on the team, and more about how the members of the team respected each other and worked together.

7.3.3 Confidence in the abilities of team members

If one sets out expectations and appropriately empowers team members whilst indicating belief in their capabilities, then there is a much higher probability that the team members will self-motivate and actually deliver. Instead of attempting to exert control over people, maybe it would be better to empower people to do the job you are asking them to do to the best of their capabilities. Do not worry too much about the ones who do not act responsibly, these will be detected soon enough. Healthy leaders choose to be driven by their faith in people not by assumptions that workers are intrinsically slackers and require control, discipline, carrots and sticks.

7.3.4 Belief in the power of positive thinking

Positive psychology working environments make for more creative, safer, fruitful teams and a more motivated and engaged workforce. Positive thinking drives success! On the other hand, disrespect leads to a reduction in cognitive ability which leads team members to make more mistakes. Leaders must project a positive attitude. Healthy leaders will hold themselves responsible for safeguarding these principles and give an example with their own behaviour.

7.3.5 Recognition of the importance of empowerment and trust

Healthcare organizations are very complex environments, Medical Physics a complex area. It is important to recognize this and to accept that it is rarely possible as a leader to have all the necessary knowledge and skills—one will therefore be dependent on others. This is an opportunity to empower the members of one's team so that they have the opportunity to exercise and develop personal initiative. There are two approaches when faced with a problem: take the top-down approach and try to come out with a solution which is delivered to the team or present the problem to the team and together develop a response. Sometimes there is no time to engage with the team and some members of the team would be tasked to solve the problem themselves without supervision. The question of trust arises. Without trust both mono-disciplinary and multi-disciplinary teams would simply end up dysfunctional. The problem arises: how does one recognize trustworthy people?—unfortunately, such people do not come with a label on their forehead reading 'I'm trustworthy', in fact it is often the reverse: they hide their faults behind a façade of flowery language interspersed with words like 'ethical' and 'effective'. Basically experience teaches one how to avoid toxic people, bluffers and those who out of personal frustrations seek to disrupt the achievements of others.

7.3.6 Acknowledgement of the importance of connecting the team to the world outside

A good leader would use all available opportunities to connect the team to the wider world—networking at both the national and international level has become an essential ingredient for success. The breadth of knowledge and array of skills necessary to deliver a project or achieve an objective in today's fast developing world means that even the collective knowledge and skills of a group may not be sufficient and persons external to the group would need to be consulted. It is important for any team to feel that they are part of something bigger than themselves and that there are others out there who would be ready to help if necessary.

7.3.7 Appreciation that developing a motivated team needs hard work

A highly functional, self-motivated, happy team is rarely the result of chance—it is usually the result of a successful leader who has reflected deeply and worked very hard to produce such a team. Producing a good team means having quality time for individual team members and for the members collectively as a team. The leader needs to read about and analyse personal psychologies, motivations and team dynamics on an ongoing basis.

7.3.8 Commitment to developing the leadership qualities of team members

As discussed in chapter 1, one of the hallmarks of a good leader is to *develop more leaders.* This is even more important today when the number of people who are willing to assume the responsibilities of leadership is low. A good leader will recognize those members of the team with leadership potential, encourage them to take on responsibilities and mentor them in their development as leaders.

7.3.9 Stretching without overstretching

A team which is not stretched is a team that does not develop; the wish to reach higher personal levels of attainment is a very strong motivator and gives a sense of achievement and a psychological buzz. However, setting goals which are far above the actual capabilities of the team leads to disappointment, demotivation and a sense of failure. A good leader would set high goals but before they do so they ensure that team members are prepared for it. The team members would know that they will create the conditions and make available the resources necessary for the individual members of the team to develop themselves to a point where they would be confident of achieving the targeted objectives.

7.3.10 Preservation of open communication channels

A good leader keeps communication channels open at all times. They know that ongoing communication is what binds the team together and what binds the team to them. The door of their office is always open for advice and for prompt decision making. No member of the team should end up feeling isolated. Each member of the team should be confident that a willing ear is available at all times. Open communication channels will help to avoid misunderstandings both at the personal and professional levels. Misunderstandings are often sorted out by a short discussion at the right time. If not tackled and nipped in the bud misunderstandings sometimes grow to a point when they are no longer manageable.

7.3.11 Honesty and integrity

Honest hard-working people follow honest leaders. If a leader is dishonest, honest people leave and experience shows that ultimately it is the honest people who deliver—you lose them, you lose everything. You cannot expect others to be honest if you are not genuinely honest yourself. Dishonesty can be camouflaged for a short time but camouflaging dishonestly is not sustainable—you will ultimately be found out.

7.3.12 Serving with empathy and humility

Finally, and perhaps most important of all, healthy leadership is based on empathy and humility.

7.4 What is leadership style?

Leadership style is the attitude the leader adopts when providing direction, formulating and implementing proposals, and motivating the members of the group. It is the 'social climate' created by the leader *as experienced by the members of the team* and includes the overall pattern of explicit and implicit actions exhibited by the leader [3]. Group culture refers to the values, beliefs, norms and objectives of the group. A pleasant group culture leads to high levels of motivation and loyalty. The main factors affecting group culture are the workplace environment, the network of communications across the group and *very importantly the philosophies and styles of the leadership. If the leaders are strong the group culture would be mainly determined by their leadership style.*

The first major study of different leadership styles was performed in 1939 [4]. The study established the three basic leadership styles:

- *autocratic*: the leader tells team members what to do and how to do it, without asking for their advice, team members expected to follow instructions;
- *democratic*: the leader includes team members in the decision making process, but the leader maintains the final decision making;
- *laissez faire*: the leader permits team members to make the decisions without interference, however, the leader is still responsible for the decisions that are made.

Today we consider more styles of leadership than these basic three and a compendium is given below. In practice, there is no single ideal universal leadership style which is superior to all others and which is applicable in all circumstances—a strategic leader would be capable of 'wearing' different styles. Each style has its advantages and disadvantages and the style to be adopted would be a function of the characteristics of the particular group and specific objectives to be achieved. The position of the leader within the organization or the stage of one's career may also influence the leadership style adopted. What is important is to be flexible and reflect on the best leadership style or combination of styles to adopt for the particular situation at a given point in time. It is also very important to consider your own personal approach to relationships in general as adopting styles which are totally alien to one's own philosophy of life and attitude towards relationships is psychologically taxing and unless the leader is capable of transforming themselves and adapting to the style of leadership that they choose they will ultimately fail.

7.5 A compendium of leadership styles

A leader's effectiveness rests primarily on her/his ability to read a situation and skilfully adopt a leadershp style that meets the individual and collective needs of the group or team and the task at hand. A compendium of leadership styles follows. Consider the advantages and disadvantages of each and possible circumstances when adopting the style would be advisable.

7.5.1 Autocratic (also known as authoritarian) leadership

This type of leadership may be required when one needs to focus almost entirely on results and efficiency. These leaders often make decisions alone or with a small, trusted group and expect employees simply to follow instructions. This style can be useful in organizations with strict guidelines, standard operating procedures or when compliance with law or standards is required. It is also important in the case of inexperienced or unmotivated groups who require a high level of supervision. It should not be used when group members need to exhibit creativity or personal initiative for outcomes to be achieved.

7.5.2 Democratic (also known as participative or collaborative) leadership

A democratic leader maintains the final decision-making for himself but requests team input and feedback on an ongoing basis. A democratic style of leadership is often associated with higher levels of team commitment and contentment as group members feel that their opinions are respected. Participation and discussion are the hallmarks of such groups, characteristics which are essential when creativity and innovation are required for successful outcomes.

7.5.3 Laissez-faire (also known as hands-off or delegative) leadership

This leadership style is often adopted by leaders who need to have time to dedicate to other matters. It is the direct opposite of autocratic leadership and focusses on the delegation of tasks to team members whilst providing minimum to no supervision. To be successful such a leadership style requires highly experienced team members who are aware of their strengths and limitations and are capable of respecting the boundaries of others.

7.5.4 Bureaucratic (also known as organizational) leadership

Bureaucratic leaders expect their team members to follow predetermined documented rules and procedures. There is an emphasis on fixed duties within a hierarchy where each team member has a well-established list of duties and responsibilities with little need for collaboration and creativity. This style is often the norm in highly regulated departments such as healthcare or finance.

7.5.5 Visionary leadership

Visionary leadership is inspirational and is essential for all groups big and small when major changes are on the agenda. Visionary leaders motivate group members by their ability to develop vision and their ability to convince and motivate group members to strive hard to achieve that vision.

7.5.6 Mentoring (also known as empowering or coaching) leadership

A mentoring leader helps each individual member of the team develop their own strengths and eliminate weaknesses. They analyse the motivations of each team member, assign appropriate mutually agreed challenging tasks, discuss expectations and provide regular feedback. Positive thinking and empowerment are the hallmarks of this leadership style which is conducive to producing other leaders, hence ensuring that the group can expand its activity without overtaxing the leader. It also ensures continuity in the development of the group should the leader be indisposed. Its main disadvantage is that it is initially very time-intensive.

7.5.7 Servant leadership

Servant leaders believe that the best way of motivating the group is by dedicating their time and expertise to the well-being of the members of the group. They believe that happy groups with a high level of morale at both the individual and collective level means better output. Such a leadership style is very important when retention of staff is important and when group members need to interact with clients.

7.5.8 Empathic leadership

Empathic leaders excel in the skill of recognizing and understanding the feelings and perspectives of individual members of the team. This skill facilitates engagement of the leader with the members of the team and helps team building. However, if not tempered with the voice of reason it can impede good decision making and cloud good judgement.

7.5.9 Ethical (also known as moral) leadership

This style of leadership is characterised by an emphasis on moral/ethical behaviour, the setting of good example and the holding of members of the team to the same standard.

7.5.10 Pacesetter leadership

Such leaders are primarily focussed on results and performance; they set high standards and expect team members to achieve their goals. This leadership style is useful in situations needing a fast-paced setting. Again it requires members with a degree of experience and needing less supervision or coaching.

7.5.11 Transformational leadership

The transformational leader is driven more by a commitment to organizational objectives than individual members' goals, however, there is an emphasis on quality communication, motivation of individual group members, less supervision and appropriate delegation.

7.5.12 Transactional Leadership

A transactional leader is someone who is totally focussed on performance based on predetermined incentives (monetary rewards, promotion, assignment of interesting projects, funds for travel) and punitive action for failure. Such leaders are concerned with establishing the right criteria for rewarding team members or otherwise. In the extreme it is based on an 'if you do this for me, I'll do this for you' approach. However, such extrinsic forms of motivation wear off in the long-term and reduce team members' intrinsic motivation to succeed. Transactional leaders do, however, include mentorship, instruction and training to make it possible for team members to achieve goals. Again this is a style of leadership that can work when the specific goals that need to be achieved do not require a high level of creativity.

7.5.13 Facilitative leadership

The facilitative leader is one who believes that it is the group that ultimately produces the outcomes, and his role is to facilitate the process for the team members, hence improving productivity. The facilitative leader uses group facilitation skills to help teams work more effectively together and mentors individuals to help solve individual problems.

7.5.14 Self-reflection/discussion point

Study each of the situations given below. Which leadership style is being adopted or should be adopted? What would you do to improve the leadership situation and make it healthier?

- Before an evaluation visit by members of the national radiation protection authority, John the lead Medical Physicist of a large Medical Physics department carefully goes over the rules and processes of the department with the leads in the various specialties. Some areas of weakness that have been identified need to be brought up to scratch by the date of the visit which is not too far off. He also holds a meeting for all the members of the department to ensure everyone is clear on the objectives of the visit and his expectations so that the inspection goes as smoothly as possible.

- As a lead Medical Physicist, Alberto has hired several bright and focusses team members whom he trusts. The population of the catchment area of the hospital has increased dramatically over the last few years and the department needs to expand its facilities and hire new staff. Alberto allows the leads in the various specialties to set up a detailed report on what needs to be done in each specialty. He only acts as the final arbiter for the team before he gives them the go-ahead to implement their plans. He is there to present possible improvements for the consideration of the specialty team leaders and addresses any questions and issues to facilitate implementation.
- Maria is the head of a large Medical Physics department in a major academic teaching hospital with a wide research focus and with a wealth of research projects. Maria spends a substantial part of her energies in applying for research grants and liaising with the heads of other professions within the multidisciplinary research teams. When welcoming new recruits to the department who have been carefully selected based on previous research experience and publications, Maria explains her role and the fact that all new researchers are free to set their own research agendas and follow their own work timetables as long as these dovetail seamlessly with the research areas of focus of the department and its cooperating departments.
- The new head of a small Medical Physics department makes a list of the tasks that they consider necessary to be done. They email each member individually and present each with a list of tasks they are expected to achieve over the next three-month period.
- The head of a large clinical Medical Physics department gathers their team leaders every six months for a meeting to discuss the experiences and issues arising from the previous six-month period. They have previously asked the team leaders to carry out a SWOT analysis for their individual teams with respect to their strategic targets and tasks. The head then praises specific team leaders for outstanding performance and goes over their achievements. They then berate underachieving teams in front of the others.
- A new head of a diagnostic Medical Physics department arrives at the department and to their consternation finds a team who consider their role simply to be routine device quality control plus the establishing of diagnostic reference levels. This was the situation established by the previous head and they are not motivated or even willing to consider alternative scenarios.
- A new president of a national association of Medical Physicists has been elected on a platform of making the association more relevant to its members and better known among its stakeholders and society at large.

7.6 Use of the various leadership styles in the various stages of project team development

Together with the availability of financial resources, team development has the strongest effect for improving organizational performance [5]. Team development in general consists of five stages namely: forming, storming, norming, performing, and adjourning:

- forming: the team is being formed, goals are being defined, members do not know each other, a proper understanding of roles and responsibilities of each is lacking. In order to avoid confusion and disorganization this stage might require a leadership style with a significant autocratic component;
- storming: the team members are getting to know each other and need to understand and perhaps adjust their own role and responsibilities and how they relate to the roles and responsibilities of other members of the team. The leadership style should take a more mentorship slant;
- norming: this is the stage where the different team members have understood their role and those of others. They start working together and may need to adjust their own personal behaviours for the good of the team. Team leaders now tend to adopt democratic and facilitative leadership styles;
- performing: in this stage the group has become a highly performing team. Team members are connected and depend on each other. Each member of the team knows the expectations of the others. A more laissez-faire leadership style is recommended as the team members require space to develop their ideas whilst requiring much less, if any direction;
- adjourning: once the project has been completed the members of the team may be considering joining another team for a new project. A laissez-faire approach would allow the different members of the team to work with others with whom they have got along well. Allowing the members to decide on their new team members may make it easier for the new team to come together [6, 7].

The members of the team will exhibit different behaviours at each stage and this often means that a different style of leadership would be required. The factor weighing most heavily on the leader's mind is how much freedom to give to the team at each stage. Too little will hamper personal initiative and possibly stifle innovation, too much and target deadlines may fall by the wayside.

References

[1] Cameron K, Mora C, Leutscher T and Margaret Calarco M 2011 Effects of positive practices on organizational effectiveness *J. Appl. Behav. Sci.* **47** 266–308

[2] Luenendonk M *The Google Way of Motivating Employees* https://cleverism.com/google-way-motivating-employees/ [accessed 23 November 2019]

[3] Newstrom J W and Davis K 1993 *Organizational Behaviour: Human Behaviour at Work* (New York: McGraw-Hill)

[4] Lewin K, Lippit R and White R K 1939 Patterns of aggressive behaviour in experimentally created social climates *J. Soc. Psychol.* **10** 271–301

[5] Macy B A and Izumi H 1993 Organizational change, design and work innovation: A meta-analysis of 131 North American field experiments, 1961–1991 ed W Pasmore and R Woodman *Research in Organizational Change and Development* (Greenwich, CT: JAI), pp 235–313
[6] Tuckman B W 1965 Developmental sequence in small groups *Psychol. Bull.* **63** 384–99
[7] Tuckman B W and Jensen M A C 1977 Stages of small-group development revisited *Group Organ. Stud.* **2** 419–27

Chapter 8

Organizational psychology (also known as occupational psychology)

Learning outcomes

By the end of the chapter the reader will be able to:

- explain the meaning of the term 'organizational psychology' and discuss its basic principles;
- discuss how a good knowledge of organizational psychology would be of help to the Medical Physics leader.

8.1 What is organizational psychology?

Organizational psychology refers to the study of human behaviour and attitudes related to the work environment. In organizational psychology, psychological theories and principles are applied to understand and if necessary help modify the behaviour of groups and individuals within the organization so that they may contribute more to the organization's success by improving performance, job-satisfaction, motivation and well-being. Organizational psychology also helps organizations and their employees handle periods of organization development and transformation. As a senior member of staff who needs to deal with people, build teams and navigate the often rough waters of the healthcare system the Medical Physics leader needs knowledge of organizational psychology.

8.2 Principles of organization psychology

The principles of organizational psychology most relevant for leadership are:

- every worker is an individual with own psychology, needs, expectations and biography. Differences in knowledge and skills, work behaviors,

doi:10.1088/978-0-7503-1395-7ch8

performance, motivation, or leadership potential can only be understood and addressed by consideration of these individual differences;
- individual performance is a function of many variables, e.g. a person with low knowledge yet high motivation can produce more results than a highly knowledgeable person with low motivation;
- individual human emotions and inter-personal relationships affect group performance;
- group dynamics affect individual performance.

8.3 Organizational psychology can help the Medical Physics leader in many tasks

Apart from the practice of leadership itself, organizational psychology can also help the Medical Physics leader in many potential leadership tasks. The table below summarises these tasks, their use in Medical Physics leadership and research techniques utilized in their application.

Leadership task	Usefulness to Medical Physics leadership	Research techniques
Job analysis	Personnel selection, performance appraisal, development of training programmes, designing standard operating procedures	Task analysis, direct observation, interviews, questionnaires, document analysis of published knowledge, skills, and competences inventories
Recruitment and selection	Identify suitable candidates from outside or from within the organization for specific tasks or roles and persuading them to apply for specific posts within the team, choosing suitable candidates for promotion	development of job announcements, definition of qualification requirements and personal characteristics, screening tools (cognitive/physical/psychomotor abilities, personality, knowledge/skills tests), simulations, structured interviews, biographical data collection, portfolio analysis
Training and training evaluation	Teaching and assessment of knowledge, skills and competences (including attitudes) for effective, safe and efficient performance in a particular work environment	Training needs analysis involving organizational analysis (goals, resources, environment), task analysis, person analysis, job analysis; instructional design and evaluation; formative and summative assessment procedures
Remuneration, compensation and benefits (tangible and intangible)	Setting of salary, bonus and benefit levels to boost motivation and performance	Comparative remuneration, compensation, benefits studies

Motivation (arousal, direction, intensity, persistence)	Identification of what promotes high performance from individual employees and teams. Should include not only salary, bonus and benefit levels but also other factors such as the physical environment, psychological climate and intangibles such as level of appreciation of work done	observation, focus groups, and in-depth interviews
Performance assessment and management	Measure performance and contrast with expectations of management—useful for giving feedback, promotion, improve training	Direct observation, performance rating scale based on job analysis, general mental ability, conscientiousness and emotional intelligence
Job satisfaction	Improve organizational commitment, job involvement, staff turnover, absenteeism	Direct observation, focus groups, and in-depth interviews
Productive work behaviour[a]	Increase technical, soft skill and for team-benefit (as opposed to self-interest) employee behaviour that contributes positively to team goals	Measurement of job performance and team-benefit through direct observation, supervisor reports, peer group assessment
Counterproductive work behaviour	Reduce factors like ineffective job performance, absenteeism, job turnover, accidents, sabotaging the work of other employees and others that contribute negatively to organization goals	Measurement of job performance and team-benefit through direct observation, supervisor reports, peer group assessment
Occupational stress, safety and well-being	Reducing loss of performance due to employee health and safety issues particularly those arising from the work environment itself including burnout	Stress audits, employee opinion surveys, risk assessments, employee absenteeism and turnover data.
Organizational culture and departmental subcultures[b]	These are intangibles that impact employee satisfaction which in turn impacts motivation and employee absenteeism and turnover	Direct observation, focus groups, in-depth interviews
Group and team effectiveness and composition[c]	Improve team effectiveness, allocation of tasks, resources, and training, leader–employee fit, ensuring individual goals and team goals are coordinated	Direct observation, focus groups, in-depth interviews

(*Continued*)

(*Continued*)

Leadership task	Usefulness to Medical Physics leadership	Research techniques
Management of organizational change	Help the teams build their self-confidence in accepting and adapting to change within the organization, hence avoiding reduced effectiveness	Periodic surveys of employee feedback regarding feelings and perspectives during the change process
Workplace bullying, aggression and violence	Avoid reduced employee well-being and team performance	Periodic confidential/anonymous surveys of employee feedback regarding workplace bullying, aggression and violence

[a] This is an umbrella term which includes job performance aspects which are directly within the control of the employee, organizational citizenship (i.e. non-rewarded behaviours such as altruism, courtesy, conscientiousness, sportsmanship, civic virtue) and innovation (creativity related behaviours).
[b] This includes shared values such as loyalty and client service, perceived level of employee support.
[c] Team composition factors include diversity of skills, compatibility in terms of personality traits/work styles/values, behaviour and dynamics.

Further reading

The Society for Industrial and Organizational Psychology (SIOP) provides many useful white papers on their website[1].

[1] https://www.siop.org/Research-Publications/SIOP-White-Papers.

Chapter 9

Organizational politics—learning to play the political game

Learning outcomes

By the end of the chapter the reader will be able to:

- explain the meaning of the term 'organizational politics' (also known as 'office or workplace politics');
- understand that organizational politics is inevitable and that trying to avoid it is bad for the group/team;
- explain the fundamentals of political games (game board, chess pieces and the rules of the game) and learn to survive them;
- acknowledge the existence of corporate bullying, toxic personalities, manipulation and office gossip.

9.1 What is meant by organizational politics?

In the most general sense organizational politics refers to the use of influence tactics to improve organizational or personal interests. Influence tactics include use of formal authority, formal and informal power, social networking and control over the flow of information (information is power!). In a more negative sense the term refers to the pursuit of individual agendas without regard to their effect on the collective well-being of the organization as a whole. However, it is to be emphasized that organizational politics may have simultaneous benefits for both organization and individuals and benefit to one does not necessarily imply loss to the other. For example, the formation of interpersonal relationships may not only lead to personal advantage but also increase efficiency and help accelerate the change processes within the organization. However, on the other hand, organizational politics focussing exclusively on personal gain at the expense of organizational well-being

doi:10.1088/978-0-7503-1395-7ch9

can destroy an organization from within. In particular, organizational politics may hinder the free flow of information within the organization, strategy development, budget setting, allocation of human resources and leadership—all factors which are critical for success. The degree of politicization of an organization varies from one organization to the next. Both individuals and subgroups within an organization involve themselves in organizational politics. The causes of organizational politics are many, but mainly they are competing visions, scarce resources, time pressures, individual motivations and social and structural inequalities within the organization. Organizational politics are an inescapable fact of life in every place of work and trying to avoid them is a delusion and waste of time and energy. Individuals with good political skills have a greater impact on organizational outcomes and often do better in acquiring more personal power, managing work related stress and job demands. *It is therefore important for Medical Physics leaders not to be naïve but to make efforts and take time to understand the forms organizational politics can take in their own organization and how to use them for the well-being of the organization and their team. One must learn to steer successfully through the political maze of one's organization, use politics for the happiness of all and help keep dysfunctional politics under control. The real challenge is how to use organizational politics for simultaneous organization and personal gain without compromising one's own personal integrity or taking unfair advantage of others.*

9.2 Medical Physicists and the political game

If Medical Physicists are to survive in the complex political environments of healthcare organizations, they must give the politics the quality of attention they give to their scientific work. However, for Medical Physicists politics is often abhorrent—frankly most of us hate it and with good reason—as it detracts from our more interesting and satisfying scientific and clinical work which we are passionate about; we only engage in it because there is really no alternative. *The only way that a scientist can accept to engage in organizational politics is to look upon it as a grand cerebral board game of sorts (think chess!). In order to become good at the game one should consider the study of the politics of the organization of which he/she forms part as a form of an observational scientific enquiry.*

Organizational politics is a game that involves a benefit versus risk assessment at each move. If one is to win, political games must be played with alertness and a full understanding of the game board, chess pieces and game rules. In order to become a good player and win you need to study:

- the game board—the various political 'terrains' in which the game is played;
- the chess pieces—the political players involved;
- the rules of the game.

Only then can one be able to assess benefits and risks, probabilities of success, wise choice of skirmishes to engage in, what to avoid, how to make the most of one's strengths, how to conceal weaknesses and choose successful combat strategies. One certain way to lose the game before you even start is by not knowing the political terrain, players and rules. *A word of warning—playing the game of organizational politics may be mentally exhausting like any chess game and one should prepare oneself psychologically for it. But don't worry, with experience it becomes easier and less taxing, you learn how to recognize the good guys who can be your allies and smell the bad guys from a mile away.*

9.3 The game board terrain

Jarrett [1] describes four types of political terrain to be found on the political game board where political strategies are played out, which he metaphorically calls the High Ground, the Rocks, the Woods and the Weeds. The four types of terrain can be presented in matrix form (see below) according to whether the activities are at the organizational/individual level or whether the source of power is formal/informal.

		Source of power	
		Formal	Informal
Level at which political activity takes place	Organizational	High Ground	Woods
	Individual	Rocks	Weeds

9.3.1 The High Ground

This refers to the *formal reporting structure* agreed to during the setting up of the organization by the Board of Governance of the organization. It describes the structure of the organization and is often illustrated through the use of a formal *organizational chart*. Organizational hierarchies reflect organizational objectives, size of the organization, quantity of resources available and the types of leaders within the organization. The High Ground ensures that the organization remains compliant with legal demands and protects the organization from excessive individual level politics. However, the High Ground can also become too rigid and bureaucratic and hinder change processes.

Several kinds of organizational entities and structures can be found within a hospital's organizational chart:

- DIVISIONS: Most large hospital organizations are structured in terms of self-contained divisions with each division focussing on delivering a single medical service. Each division would include its own BACK-OFFICE SUPPORT GROUPS e.g. human resources, legal department, education and training, IT people, *Medical Physics*. This is the case when hospital divisions such as D&IR, NM and RO employ their own Medical Physicists independently and when an independent Medical Physics department may not exist. As with all things in life, the structure has its advantages and disadvantages with respect to the position of Medical Physicists. In the case of forward-looking divisions with enlightened heads, the arrangement has many advantages as Medical Physicists can interact very closely with the other healthcare professionals involved in the Division and have a major impact on service quality. Their role may be much appreciated. On the other hand, a head who does not appreciate the importance of the role or has an interest in camouflaging inadequate dose monitoring or dose optimization or simply wants to cut costs indiscriminately would find it easy to eliminate Medical Physicists whose job it is to ensure that patients are protected from unnecessary radiation.

- INDEPENDENT SUPPORT GROUPS: some hospital organizations have *independent* Support Groups (e.g. independent Medical Physics departments) with a separate budget which service the various Divisions. Since such Support Groups are intrinsically back-office affairs; they do not have a direct revenue stream from clients (in this case patients or health insurance companies). In the case of revenue dominated hospitals run by economics-oriented CEOs with little or no direct healthcare knowledge they often need to constantly justify their existence and their role is poorly understood by management. This creates a certain political tension. *Medical Physics leaders running independent Medical Physics divisions need to combat such attitudes by publicizing what they do in an ongoing proactive manner.*
- Increasingly, large hospitals have a Division *by* Support Group MATRIX ORGANIZATIONAL STRUCTURE. Resources from each Support Group are *temporarily* assigned to particular Divisions on a need-to basis. The relation between the Support Groups and Divisions is marked with a dotted line on the organizational chart and hence this is often referred to as dotted-line reporting. There is a constant tension between matrix managers and Division/Support Group managers. Problems usually arise over the sharing of common Support Group resources.
- FLAT ORGANIZATIONAL STRUCTURES: On the other hand, small clinics can consist of a single multi-service 'division' with a flat reporting structure with minimal hierarchy. Such groups may work in more open-plan type environments so that workflow is visible to all.

Many organizations have hybrid structures and the political landscape gets another twist. In systems which include lots of executives there is often a lot of infighting as executives fight over influence and resources.

9.3.2 The Rocks

This refers to the *way formal power is exercised in practice by authoritative individuals within the formal organizational structure*. Such individual formal power arises from title, role, access to resources, membership of high status groups like the senior management team, financial committees or influential task groups. The appropriate exercise of personal formal power by individuals is essential to the well-being of the organization and the personal happiness of the individuals within that organization; however, inappropriate exercise of personal power by individuals can produce excessive conflict within the organization with subsequent negative effects on organizational goals.

9.3.3 The Woods

The Woods refers to the *informal, implicit, spoken/unspoken ecosystem of norms, values, protocols, routines, habits, axioms, assumptions which permeate activities across the organization. The Woods can be described as the organizational culture.* The Woods can be a very good influence in organizations based on an emphasis on ethical behaviour with respect to patient health and mutual respect between professions. *Medical Physics with its emphasis on quality and safety thrives in such environments. However, in organizations where austerity and cost-cutting becomes the dominant 'value' Medical Physics might suffer.*

9.3.4 The Weeds

This refers to the *informal personal networks and hierarchies running parallel to the formal organizational structure* that lead to personal influence. Such informal networks can lead to high influence, e.g. being on good terms with the personal secretary of a senior manager (and who controls their agenda) can get you quick appointments not available to others. It is important to know where you fit in the Weeds. It is important to get to know these key players as they are a source of informal power and information (and information is power!). Make use of them when necessary and stop them in their tracks and isolate them when harmful (e.g. by strengthening connections with other key players or by developing and disseminating a counter-narrative to what they represent). It is perhaps also important to identify any key gaps in this informal network which you can fill yourself hence increasing your influence. Jarrett calls these networks the Weeds because they grow naturally without the need for any intervention, however, weeds if allowed to grow unchecked can undermine an organization.

9.4 The chess pieces—the political players

Like in a chess game there are several types of chess pieces (political players) in the game of organizational politics. As in chess we have several role players each with their part in the game. Some people may assume several roles or change roles according to the state of play. Here are some of the ones you may meet:

The CEO: this is the person with the maximum decision making power and who everybody tries to be on good terms with.

The vice-CEO: this person is being prepared to become CEO, helps the CEO in tasks which the CEO has no time for, has a lot of political power and often is a gatekeeper to the CEO.

The Empire Builder: increases their political power by employing more and more people in their division even when not really necessary. Their ultimate aim is usually to become CEO.

The Yes Man: always agrees with the CEO or the vice-CEO even when they are aware that the CEO is making a mistake.

The Curmudgeon: always tries to belittle the achievements, arguments of others hence trying to give the impression that they would do better.

The Vortex: this is the drama queen attracting attention to themselves and their own agendas.

The Peacemaker: they try to get people to work together and reduce the level of conflict within the organization. They really have the interests of the organization at heart.

The Fake-Peacemaker: this one plays the peacemaker, however, it is only a smoke-screen to hide their real intentions or get others to lower their defenses.

The Brain: this guy is data/evidence-driven. May go overboard and simply present data to show how smart they are even when the data is irrelevant to the issue at hand.

The Rubber Chicken: makes a lot of noise but often just for the sake of attention seeking, their interventions may be of little relevance to the issue at hand.

The Parrot: often just parrots what they perceive to be the idea that is currently winning the day.

The Thief: steals the ideas of others and presents them as their own.

The Corporate Bully: bullying in the workplace and on boards can be verbal, nonverbal or psychological. It can be carried out by superiors, peers or even subordinates. Workplace bullying often happens within the normal rules and policies of the organization so it may not be so recognisable.

9.5 The rules of the game

Here are some rules for organizational games:

Going through the proper channels: this essentially means that at all times decisions should be taken by the relevant person at the appropriate level in the formal organizational chart. Apply rule only when 'the relevant person at the appropriate level in the formal organizational chart' is a person of goodwill. If otherwise, circumvent if possible.

Knowledge is power: information is a very precious commodity within an organization and having access to it and preventing access to opponents gives advantage in the political game. Make sure that your opponents are not blocking your information channels.

No bad news: nobody likes bad news even if it is true. Learn to break bad news in a positive way and if possible suggesting possible solutions, otherwise you will be seen negatively.

Divide and conquer: one of the ways of controlling your enemies is to make sure that they never band up against you. Do not allow them to split your team or separate you from your allies.

Favourites of the CEO are untouchable: if the favourites are honest it is a plus to all people of goodwill. If otherwise, circumvent if possible.

Favouritism wins over ability and performance when the CEO is weak: A weak CEO can only survive by pandering to their loyal acolytes. Keep a low profile, let the acolytes destroy each other.

Cover yourself from subterfuge by communicating openly what's going on: one means of defense is by ensuring everybody knows when you are under attack so that others are not unknowingly manipulated into joining your attackers.

If you want something done badly, do it yourself: in organizations with a high level of political gaming you are the only one that you can trust.

If you want something to fail ask for a committee: this is the reverse of the previous rule—ask for a detailed analysis, a large committee (impossible for them to find common meeting times), suggest an overworked person as chairperson.

Informal networks speed things up: to be used when you know that the formal network is bound to fail or be too slow.

Use your alliances: in any strategy game, alliances will help keep you on top and defend you in times of difficulty.

Data/evidence-driven squashes politics: when you want to reduce the effect of politics insist on data/evidence based decisions.

Going through kingpins: know who the *real* kingpins in the organization are (usually you find them as heads of important committees like the financial committee).

Have a plan B: include 'what ifs' in your strategies.

The prevailing leadership style in the organization is dominant: assess what the present predominant leadership style of the organization is and work within it or around it, do not oppose it unless a critical mass of opponents has been achieved.

Be quick to adapt to changing circumstances: you need to be ready to change your strategy fast, do not be enamoured of particular strategies if you see that they are not working. Remember it is not the most intelligent who win but those who are most adaptable.

Fight or flight: keep this under control. Choose which battles are worth the fight. If it is a crucial issue flight is not an option, at most a temporary retreat.

Stand up to the corporate bullies: as with all bullies, if you give in to them, they will dominate you.

It is not winner takes all: get as much as is possible in the circumstances.

9.6 Surviving and winning at political gaming

In this section we give some general advice to help you be more successful in organizational politics, warn you about toxic personalities, discuss the issue of manipulation and consider office gossip and its effect on the flow of information within an organization.

9.6.1 Some general advice

Be positive.
Be prepared.
Be data/evidence driven.
Learn to observe and listen.
Foster alliances.
Admit when you are wrong.
Understand the real political issue behind the topic under discussion.
Tell the truth.
Avoid communicating ambiguous proposals in writing (avoid emailing them!!).

Put the interest of the patient first.
Stand up for yourself calmly but firmly when necessary.
Help others.
Try and find common ground.
Agree to disagree.
Be a peacemaker.
If you do not know something just say so.
Do not allow yourself to be provoked.
Never raise your voice in anger.
Know your trigger words.
Attack only when absolutely necessary and when conditions are in your favour.
Know when to keep silent and when to retreat and regroup.
Let what is most important for you guide your actions.

9.6.2 Recognizing and dealing with toxic players

The three dark personality traits, narcissism, Machiavellianism and psychopathy, have central significance in understanding the damaging side of organizational politics. Narcissism is characterized by egotism and lack of empathy; Machiavellianism by manipulation, deception and exploitation of others; psychopathy by antisocial behaviour, selfishness, cruelty, and pitilessness. You find such people in all areas of social activity and healthcare organizations are no exception. *Read about these personality traits, learn to recognize them and how to deal with them—ignore them at your peril, forewarned is forearmed!*

9.6.3 The issue of manipulation

Manipulation is when individuals or groups use underhanded means to achieve their own personal objectives at the expense of other individuals or groups who may be more capable. This is the strategy often chosen by those whose ambition is much higher than their ability. For example, selection for promotion, should be made purely on merit. If a selfish Medical Physicist believes that they would lose in a fair competition, they may use coercive means of influence and intimidation to push themselves forward. A professional group that may see Medical Physicists as stronger than them would avoid fair competition rules. Manipulators often co-opt as many colleagues as possible into their plans as allying with unsuspecting others strengthens their personal position. Such individuals and groups consume time and resources for their own gain at the expense of organizational goals. *Read about manipulation, learn to recognize it and how to deal with it—ignore it at your peril, forewarned is forearmed!*

9.6.4 Office gossip and the flow of information within an organization

Unscrupulous organizational politics differs from office gossip in that people participating in manipulative organizational politics do so with the aim of gaining personal advantage, whereas gossip can be a purely social activity. However, the two are often connected and office gossip is one of the techniques used by manipulators to control the flow and content of information, hence gaining advantage. Information can be distorted, misdirected, or blocked in order to create doubt on the intentions of competitors. *Read about office gossip, learn to recognize it and how to deal with it—ignore it at your peril, forewarned is forearmed!*

9.7 Final words of advice

In times of intra-organizational struggles, data/evidence driven employees who construct their political strategies on safeguarding the best interests of the organization and its clients (in our case the patient) will have an easier time diffusing political conflicts and ensuring that their honourable motivations will be unquestioned. In the long run honesty and integrity is recognized by all. One has only one life, choosing to live by an honour code is the way to go.

Reference

[1] Jarrett M 2017 The 4 types of organizational politics *Harvard Business Review* https://hbr.org/2017/04/the-4-types-of-organizational-politics [accessed 12 September 2019]

Chapter 10

Negotiating skills for the Medical Physics leader

Learning outcomes

By the end of the chapter the reader will be able to:

- distinguish between negotiation, mediation and arbitration;
- distinguish between distributive and integrative negotiation;
- explain how to prepare for a negotiation session;
- recognize the different types of negotiators;
- recognize the different types of negotiating tactics.

10.1 Negotiation, mediation and arbitration

Negotiation is direct discussion between two parties involved in a dispute in order to achieve consensus. On the other hand, mediation involves having an independent neutral third party to act as a catalyst to help the two sides come to a consensus. Finally, in the case of arbitration, the third party listens to the arguments of both sides and then takes the final decision themselves, both sides are bound to accept the decision of the third party. A Medical Physics leader might need to be a party when say negotiating quality control duties with radiographers in an imaging department, mediate or even arbitrate between two members of their own department over who would present research results at a conference, arbitrate as an independent referee when asked to do so by a leader of another team. This chapter deals with negotiation.

10.2 The two types of negotiation

There are two types of negotiation, distributive and integrative. In the case of DISTRIBUTIVE NEGOTIATION (also known as win–lose or zero-sum-outcome) both parties want the same piece of pie so each side adopts an extreme fixed position and each side cedes as little as possible. Essentially it's a battle of wits. INTEGRATIVE NEGOTIATION on the other hand aims for a win–win outcome.

doi:10.1088/978-0-7503-1395-7ch10

The negotiation process is not considered as a tug-of-war with parties on opposite sides, but a problem to resolve together. There are attempts to avoid problems by creating value in terms of expanding the pie, reframing the issue or compensating for a loss through alternative gains.

10.3 Productive negotiation

Negotiation, particularly distributive negotiation, is often a tiring task and can lead to acrimony and even lasting bitterness—particularly when one side sees the other party as being selfish and aggressive.

However, negotiation can be a positive process provided one keeps in mind the following rules:

- look upon negotiation as shared problem-solving not conflict resolution;
- use objective criteria;
- each side should appreciate the emotions and motivations of the other side;
- separate issues from personalities;
- explicitly discuss each other's perceptions;
- parties should not be tied rigidly to present perceptions—be ready to change if reasonable;
- apply good communication and listening skills;
- speak with a clear purpose;
- each side should do its homework—in complex settings map out all potential relevant issues and then make appropriate connections;
- in healthcare keep the patient's interest and inter-professional respect paramount—then you know that you are doing the right thing!

10.4 Preparing for the negotiation

Remember negotiation is part problem-solving, part controlling the egoistic tendencies of others (and sometime your own), part game, and part lottery. *The secret of success in negotiation is advance preparation—both content-wise and psychology-wise.*

Preparing content:

- think deeply and clearly about what you actually want, what is essential and what is merely icing on the cake;
- prepare the data and evidence to back your arguments;
- if you are part of a team, make sure you agree between you on the issues *before* facing the other side;
- if any mediator is present, make sure you enquire about his approach and attitude;
- what to do if the other party shows bad faith and only pretends to negotiate?
- VERY IMPORTANT: what are your red lines?
- VERY IMPORTANT: what is your preferred alternative if no agreement is reached?
- What questions will you be asking to assess the red lines and preferred alternative of the other party?

Psychological preparation:

- avoid taking difficult people with you, you need help not additional problems;
- think how you will be facing the negotiation in an emotionally intelligent way;
- seek a calm before the negotiation session proper;
- remind yourself that you will not allow anybody to provoke you, if you show anger you've lost it!

10.5 Types of negotiator: which type of negotiator are you—and which your opponent?

Here are the main types of negotiator. Which type are you, which type your opponent? Can you handle the negotiating style of your opponent? Should you take others with you for support? *Remember the future of your group may depend on your negotiating skills and attitude.*

The soft negotiator:

- the avoider: does not like conflict at all, avoids negotiation, skips meetings, and even leaves a meeting so as to avoid conflict;
- the accommodator: solves the problems of the other side instead of their own, seeks to preserve personal relationships at all costs, withdraws complaints rather than argue, may be taken advantage of, does not separate the people from the problem, yields to the demands of others even when unreasonable or unethical, makes concessions too quickly.

The hard negotiator

Highly competitive, tries to force others to accept their views, bullying tactics, neglects importance of relationships, challenges continuously, accuses, complains, intimidates, sets pre-conditions.

The principled negotiator

A collaborator, likes solving tough problems creatively, understands the concerns of the other side but also keeps their own in mind, tries to find solutions to satisfy both sides. *Most Medical Physicists are this type of negotiator, because we are inherently problem solvers. However, keep in mind that you may have to deal with others who may not be principled at all—in fact you may encounter types of persons whose psychology and level of intelligence does not permit principled negotiation.*

10.6 Negotiating tactics

Here are some well-known negotiating tactics you might meet when dealing with hard negotiators. Some of them verge on the unethical, *even though you would not use them yourself you still need to recognize them when others are using them so that you may counter them effectively.*

Nibble: ask for small things before closing the meeting when people are tired so that they accept without thinking things over carefully.

Snow job: flood opponents with info and jargon so that they lose track of what is important and what is not.

Bogey: make one issue appear to be important when it is not and then trade it for a real major concession.

Brinkmanship: one party pursues aggressively their demands to a point where the other party gives them what they want or simply walks away.

Short deadline: force opponent to make quick erroneous decisions.

Flinching: fake strong negative reactions (expressions of shock, surprise, gasping for air incredulity) to give the other party the impression that what they are asking for is absurd with the hope that they would back down on their demands.

Good guy/bad guy: one team makes unreasonable demands; another makes more reasonable demands to make the other look bad.

Highball/lowball: the first party takes on an extreme position which makes the second party re-evaluate their demands, after which the first party tones things down to appear reasonable.

Anchoring: gain advantage by verbally expressing a position first.

Auction: when more than one person wants the same thing, pit one against the other.

Further reading

If you want to go deeper read the latest editions of the following books:

- Bryson J M and Alston F K 2011 *Creating Your Strategic Plan: A Workbook for Public and Nonprofit Organizations* (San Francisco, CA: Jossey Bass)
- Truxillo D M, Bauer T N and Erdogan B 2015 *Psychology and Work: Perspectives on Industrial and Organizational Psychology* (New York: Psychology Press/Taylor & Francis)
- Vigoda-Gadot E and Drory A (eds) 2006 *Handbook of Organizational Politics* (Cheltenham, UK: Edward Elgar Publishing)
- Rollinson D 2008 *Organisational Behaviour and Analysis: An Integrated Approach* (Harlow, UK: Pearson Education)
- Oliver J 2013 *Office Politics: How to Thrive in a World of Lying, Backstabbing and Dirty Tricks* (London: Vermilion Press)
- McIntyre M G 2005 *Secrets to Winning at Office Politics: How to Achieve Your Goals and Increase Your Influence at Work* (New York: ST Martin's Griffin)
- Donaldson M 2007 *Fearless Negotiating* (New York: McGraw-Hill Education)

www.ingramcontent.com/pod-product-compliance
Ingram Content Group UK Ltd.
Pitfield, Milton Keynes, MK11 3LW, UK
UKHW051848100926
8556IPUK00005B/517